D0848598

Hitler's Ethic

List of Previous Publications

From Darwin to Hitler: Evolutionary Ethics, Eugenics and Racism in Germany. New York: Palgrave Macmillan, 2004.

Socialist Darwinism: Evolution in German Socialist Thought from Marx to Bernstein. San Francisco: International Scholars Publications, 1999.

The Myth of Dietrich Bonhoeffer. San Francisco: International Scholars Publications, 1997.

Hitler's Ethic

The Nazi Pursuit of Evolutionary Progress

Richard Weikart

palgrave
macmillan

First published in 2009 by
PALGRAVE MACMILLAN®
in the United States—a division of St. Martin's Press LLC,
175 Fifth Avenue, New York, NY 10010.

Where this book is distributed in the UK, Europe and the rest of the world,
this is by Palgrave Macmillan, a division of Macmillan Publishers Limited,
registered in England, company number 785998, of Houndmills,
Basingstoke, Hampshire RG21 6XS.

Palgrave Macmillan is the global academic imprint of the above companies
and has companies and representatives throughout the world.

Palgrave® and Macmillan® are registered trademarks in the United States,
the United Kingdom, Europe and other countries.

ISBN: 978–0–230–61807–7

Library of Congress Cataloging-in-Publication Data

Weikart, Richard, 1958–
    Hitler's ethic : the Nazi pursuit of evolutionary progress / Richard
Weikart.
        p. cm.
    Includes bibliographical references and index.
    ISBN 0–230–61807–3
        1. Hitler, Adolf, 1889–1945—Philosophy. 2. Hitler, Adolf, 1889–1945—
Political and social views. 3. Eugenics—Germany—History—20th century.
4. Ethics, Evolutionary. 5. National socialism—Philosophy. 6. Germany—
Ethnic relations—History—20th century. I. Title.

DD247.H5W375 2009
171'7—dc22   2009002646

A catalogue record of the book is available from the British Library.

Design by Newgen Imaging Systems (P) Ltd., Chennai, India.

First edition: August 2009

10 9 8 7 6 5 4 3 2 1

Printed in the United States of America.

CONTENTS

ILLUSTRATIONS

PREFACE

When I received my Ph.D. fourteen years ago, I never imagined that I would be writing about Hitler or Nazism. I thought the history of the Third Reich was an overworked field, so I figured the chances of uncovering anything genuinely new, interesting, and truthful were scant. Nonetheless, as I investigated the history of evolutionary ethics in pre-World War I Germany, I noticed—to my surprise—remarkable similarities between the ideas of those promoting evolutionary ethics and Hitler's worldview. This discovery (which happened around 1995) led me to investigate Hitler's worldview more closely, and this research convinced me that I had found something important to say about Hitler's ideology.

Many scholars have noticed Hitler's reliance on social Darwinism, and many recent works have explored the importance of eugenics and scientific racism in Nazi ideology and policy. I am indebted to many of these fine expositions of Nazi thought and practice, and I thank all those scholars (listed in my bibliography) for laying the groundwork for my own work.

I would also like to thank those scholars who read all or part of the manuscript: Edward Ross Dickinson, Derek Hastings, and Randy Bytwerk, as well as a couple of anonymous readers. Also special thanks to Randy Bytwerk for helping with six of the illustrations (the dust-jacket and figures 1.1, 2.1, 3.4, 5.1, and 5.2).

I am greatly indebted to my own university for providing research funds, access to Inter-Library Loan (thanks again to Julie Reuben), and a research leave for the academic year 2007–2008, which allowed me to complete the manuscript.

Lastly, I thank my precious wife, Lisa, and my seven children (Joy, John, Joseph, Miriam, Christine, Hannah, and Sarah) for their love and support.

BRIEF CHRONOLOGY

April 20, 1889	Hitler born in Braunau, Austria-Hungary
1907	Hitler moved to Vienna
1913	Hitler moved to Munich
1914	Hitler joined German military at opening of World War I
October 1918	Hitler wounded in gas attack and hospitalized
November 1918	Revolution overthrows emperor and Social Democrats proclaim a republic
1919	Hitler assigned to army propaganda unit
August 1919	Weimar Constitution approved, instituting parliamentary democracy
September 1919	Hitler joined German Workers' Party
October 1919	Hitler gave first speech to German Workers' Party
February 1920	National Socialist German Workers' Party proclaims its Twenty-Five Point Program, coauthored by Hitler
November 1923	Hitler arrested for his role in Beer Hall Putsch
1924	Hitler wrote first volume of *Mein Kampf* in prison
December 1924	Hitler released from prison
July 1925	First volume of *Mein Kampf* published
December 1926	Second volume of *Mein Kampf* published
1928	Hitler wrote his *Second Book*, which he never published
1929	Great Depression began
1930	Presidential rule began, undermining Weimar Republic
April 1932	Hitler lost presidential race to Hindenburg
July 1932	National Socialists became largest party in German parliament by winning over one-third of the votes

January 30, 1933	Hitler appointed chancellor
March 1933	Enabling Act passed, giving Hitler's cabinet emergency powers
March 1933	First official concentration camp set up at Dachau
July 1933	Compulsory sterilization law passed by Hitler's cabinet
August 1934	President Hindenburg died, and Hitler assumed his powers
March 1935	Remilitarization begun with conscription and foundation of air force
June 1935	Law allowing abortion for hereditarily ill
September 1935	Nuremberg Laws proclaimed, which discriminated against Jews
October 1935	Law for the Protection of Hereditary Health of the German People prohibits marriage of the hereditarily ill and requires a health certificate for marriage
March 1936	Remilitarization of Rhineland
November 1937	Hossbach Memorandum shows Hitler's war plans
March 1938	Austria annexed
September 1938	Munich Conference gave Hitler Sudentenland region of Czechoslovakia
November 1938	Crystal Night, first nationwide physical violence against Jews
March 1939	Germany invaded Czechoslovakia and partitioned it
August 1939	Nazi-Soviet Nonaggression Pact
September 1939	Germany invaded Poland, so Britain and France declared war
September 1939	Polish Jews forced into ghettoes
October 1939	Hitler signed decree to begin killing the disabled
April 1940	Germany invaded Denmark and Norway
May 1940	Germany invaded Netherlands, Belgium, Luxemburg, and France
June 1940	France surrendered
June 1941	Germany invaded Soviet Union
June 1941	Special units began massacring Jews in Soviet Union
December 1941	First death camp at Chelmno began systematic extermination of Jews
January 1942	Wannsee Conference organized transport of Jews to death camps

February 1943   German army defeated at Stalingrad; major turning
                point on Eastern Front
June 1944   D-Day opened up second front
April 30, 1945 Hitler committed suicide as Berlin was surrounded
May 7–8, 1945   Germany surrendered

Introduction

Why was Hitler so evil? How did he gain power in such a well-educated, civilized country? Why did so many leading scholars in Germany support Hitler's policies? These questions have perplexed and haunted humanity since the Nazi era. Myriads of historians, social scientists, journalists, psychologists, and psychiatrists have tried to provide answers. However, even after all this thoughtful reflection by multitudes of scholars, none of the explanations have proved completely satisfying. Hitler's evil still eludes our comprehension, and the reasons he committed such atrocities—with the complicity of many fellow Germans—are still shrouded in mystery. In 1998 the journalist Ron Rosenbaum dedicated an entire book, *Explaining Hitler*, to charting the many attempts by scholars to explain Hitler and his evil deeds. He concluded that we certainly have not yet explained Hitler's evil, and maybe we will never be able to explain it.[1]

Most people suppose that Hitler was a power-hungry opportunist who simply ignored morality whenever it got in his way. His lust for power overcame his conscience—if indeed he had a conscience. Some—including Rosenbaum—insist that Hitler was so demonic that he embraced evil deliberately. According to this view, Hitler was fully conscious of his immorality, but he *enjoyed* evil and exulted in it. Alternatively, some scholars portray Hitler as an amoral nihilist exerting his will to power as a Nietzschean Superman.[2] According to these typical views, any time Hitler spoke or wrote about ethics or morality, he was consciously lying (or to put a Nietzschean gloss on it, he was myth-making). He only cared about ethics and morality as a political tool to manipulate the masses and win support for himself and his regime. If this were true, it would be pointless to analyze Hitler's views on ethics and morality, unless one were studying Nazi propaganda. Though many scholars today recognize the importance of ideology in shaping Nazi policies, few have grappled with the role of ethics and morality in Nazi ideology.[3]

I will demonstrate in this book the surprising conclusion that Hitler's immorality was not the product of ignoring or rejecting ethics, but rather came from embracing a coherent—albeit pernicious—ethic. Hitler was inspired by evolutionary ethics to pursue the utopian project of improving the human race. He really was committed to deeply rooted convictions about ethics and morality that shaped his policies. Evolutionary ethics underlay or influenced almost every major feature of Nazi policy: eugenics (i.e., measures to improve human heredity, including compulsory sterilization), euthanasia, racism, population expansion, offensive warfare, and racial extermination. The drive to foster evolutionary progress—and to avoid biological degeneration—was fundamental to Hitler's ideology and policies.[4]

By embracing his particular brand of ethics, Hitler perpetrated greater evil than he would have if he had been merely opportunistic or amoral. Evolutionary ethics drove him to engage in behavior that the rest of us consider abominable. Further, even though most Germans did not entirely share his ethical views, the ethical and moral thrust of many of his speeches had tremendous propaganda effect. Those who did share his evolutionary view of ethics, such as many in the scientific and medical community, were often eager participants in Nazi atrocities.

Why, if I am right, have so few scholars noticed Hitler's ethic? First, my thesis is counterintuitive. Hitler is the epitome of immorality, so how could he have been following any kind of moral doctrine? Just suggesting that Hitler was somehow "moral" seems grotesque.[5] Second, Hitler's atrocities violated most forms of ethics familiar to historians. Hitler not only jettisoned (and at times expressed contempt for) many tenets of Christian morality, but his policies also showed disdain for Kantian ethics, utilitarianism, and most other forms of ethics prominent in Western culture. Hitler's ethic—and the policies that flowed from it—stands in stark contrast to what most people today consider moral. Finally, few historians understand evolutionary ethics, especially in the early twentieth-century racist form embraced by Hitler. Many scholars have noticed Hitler's commitment to social Darwinism, but almost no one has analyzed the ethical dimension of it.[6]

I discovered Hitler's commitment to evolutionary ethics in a roundabout way. Indeed, when I began doing research on evolutionary ethics in late nineteenth and early twentieth-century Germany, Hitler was nowhere on my radar screen. After completing my dissertation on the reception of Darwinism by German socialists in the late nineteenth and early twentieth centuries, I wanted to investigate the way that biologists and other scholars used Darwinism to formulate new views of ethics and morality in the late nineteenth and early twentieth centuries.[7] Though I had studied a good deal about Nazism in the course of my graduate studies, I had no intention of ever writing anything about Hitler or Nazism. Only as I began studying the writings of prominent biologists, eugenicists, and social thinkers who promoted evolutionary ethics did I notice the obvious parallels between their vision of ethics and Hitler's worldview. The more I investigated evolutionary ethics in late nineteenth and early twentieth-century Germany, the more I discovered the close linkages between evolutionary ethics and eugenics, euthanasia, militarism, and racial extermination. The parallels with Hitler were so evident that I expanded my original research project, ultimately publishing it as *From Darwin to Hitler: Evolutionary Ethics, Eugenics, and Racism in Germany.*[8]

Hitler's ethic was essentially an evolutionary ethic that exalted biological progress above all other moral considerations. He believed that humans were subject to immutable evolutionary laws, and nature dictated what was morally proper. Humans must adapt to and even model themselves after the laws of nature. Most important among these biological laws was the struggle for existence, as Hitler emphasized repeatedly throughout his career. Whether in public or in private, Hitler repeatedly stressed the importance of conforming to the laws of nature, especially the Darwinian law of the struggle for existence. In *Mein Kampf* he stated that humans could not escape the "iron logic of Nature," since "his action against Nature must lead to his own doom." Several pages later he commented, "Those who want to live, let them fight, and those who do not want to fight in this world of eternal struggle do not deserve to live. Even if this were hard—that is how it is!"[9] Over a decade later he told construction workers:

> It is absolutely true that first of all the law of selection exists in the world, and nature has granted the stronger and healthier the right to life. And rightly so. Nature knows no weakling or coward, it knows no beggar, etc., but rather nature knows only those who stand firm on their soil, who sacrifice their life, and indeed sacrifice it dearly, and not those who give it away. That is an eternal law of nature. You see it if you gaze into the forest, you see it in every meadow, you see it in the struggle of individual organisms

in the world, and you see it throughout the millennia of human history...[10]

In 1942—after the Holocaust was in full swing and the full significance of this deathly philosophy was more evident—he repeated this theme in a speech to army officers:

> We are all beings of nature, which—inasmuch as we can see it—only knows one harsh law, the law that gives the right of life to the stronger and takes the life of the weaker. We humans cannot exempt ourselves from this law.... On this earth we observe the unswerving struggle of living organisms with each other. One animal lives, in that it kills the other.[11]

This eternal law of struggle, in Hitler's view, produced all that was good in the world. It must continue, if further progress were to be made. Trying to escape the law of struggle would only backfire, because it would contribute to degeneration and decline.

Figure I.1 Nazi school poster: "Eradication of the Sick and Weak in Nature"

In Hitler's view, whatever promoted the health and vitality of the human species was morally good. Conversely, anything contributing to biological degeneration or decline he deemed immoral. Hitler considered mental acumen and physical prowess as the highest virtues. Those who, through no choice of their own, were mentally impaired or physically disabled were viewed as sinners, especially if they dared to propagate their "defective" heredity by bearing children. Since the Darwinian struggle for existence had produced so much biological progress, Hitler and other social Darwinists considered competition—resulting in the death of the vanquished—a positive force. At the same time, they also approved of measures to artificially improve the human species, such as compulsory sterilization of the congenitally disabled. Sharpening natural selection by increasing competition and introducing policies to artificially select humans were twin prongs of a concerted effort to foster upward evolution.

Closely linked to these notions of biological progress and degeneration was the conviction Hitler shared with other social Darwinists that humans are essentially unequal and thus need not be treated equally. For Hitler this inequality was most pronounced between different races. Except for targeting the disabled, he often downplayed the inequalities among Germans, since he wanted to cultivate national unity among those in the German People's Community (*Volksgemeinschaft*). The German term *Volk* defies easy translation, and thus many German historians, myself included, use the term rather than its somewhat anemic common translation as "people." It does indeed mean "people," but in the specific sense of an ethnic community. Nineteenth-century nationalists used it to describe a particular group of people speaking the same language and sharing a common culture. However, for Hitler—as well as for many other early twentieth-century racial thinkers—the *Volk* was defined not by cultural characteristics but by biological traits.[12] Hitler consistently used the term *Volk* as a synonym for race, which supposedly stood for biologically defined groups of people bound by common heredity.[13] The German *Volk* was almost synonymous with the Aryan race in Hitler's eyes, and his policies aimed at bringing about the political unity of all Aryans and exclusion of all non-Aryans from the German nation.[14]

Hitler believed that the so-called Aryan race was the most highly evolved form of humanity, and that because of its superior biological traits, it alone had produced all significant cultural achievements in world history. Therefore, he was convinced that whatever conferred a competitive advantage to the Aryan race in the human struggle for existence would drive human evolution to ever greater heights. Given this twisted vision of ethics and his Aryan supremacist racial views, he could morally justify all sorts of atrocities, as long as they were directed against those deemed biologically inferior.

Hitler's vision of evolutionary ethics made him contemptuous of anyone defending human rights. He frequently dismissed humanitarianism and human rights with disdain, since it would only help weaklings and destroy the biological vitality of the allegedly superior German race. In *Mein Kampf* he scorned those appealing to human rights as weaklings themselves, stating, "*No, there is only one holiest human right, and this right is at the same time the holiest obligation, to wit: to see to it that the blood is preserved pure and, by preserving the best humanity, to create the possibility of a nobler evolution of these beings.*"[15] Thus, for Hitler evolution not only trumps human rights but shows that human rights are misguided and deleterious.

In Hitler's ethic the end—evolutionary advancement—justified the means. Nothing was taboo, no matter what earlier ethical and moral systems had taught. As he explained to leaders of the armaments industry in mid-1944, Nazism was not beholden to any dogmas. Rather, he insisted, "In carrying out the struggle for life of a *Volk* there can only be one dogma, namely to apply every means that leads to success....For there can only be one single dogma, and this dogma is very briefly: The right thing is whatever is advantageous in itself, advantageous in the sense of preserving the *Volk*."[16] In other words, the end justifies the means. Survival in this human struggle for existence was the paramount virtue. This was not just a desperate cry of a dictator facing imminent defeat. As we shall see, it was a constant refrain throughout his career, though it was more candidly expressed in his private speeches and conversations than in his public speeches.

Once we understand Hitler's ethic (which I will explain in greater depth later), Hitler's morality becomes less puzzling. When I use the term "ethic" in the ensuing discussion, what I mean is a fundamental principle used to justify specific moral commands. Morality or morals, on the other hand, are specific commands or precepts. For example, many forms of Christian ethics would appeal to divine wisdom and the divine origins of morality to justify specific moral commands, such as "thou shalt not steal." Christian morality, then, would usually be based on divine revelation, so scripture, religious tradition, and one's God-given conscience would determine specific moral injunctions. Most Christians throughout history have viewed morality as fixed and universal, transcending human culture and society.

Another quite different form of ethics is utilitarianism, which determines morality by its results. Utilitarians apply a single guiding principle to judge all moral commands and legislation: the greatest happiness for the greatest number. Jeremy Bentham, the originator of utilitarianism, hoped someday humanity could calculate happiness (and thereby morality) with "felicific calculus," which would add up all the pleasures and subtract all the pains caused by any particular form of morality or legislation to determine its contribution to happiness.

Quite different from other forms of ethics, evolutionary ethics (at least most of the varieties circulating in the early twentieth century) usually justified morality by its effects on the improvement of the human species. Like utilitarianism, it judged by results. However, as Darwin himself noted in *The Descent of Man*, human social instincts produced through the struggle for existence do not aim at the greatest happiness, but rather at the survival and reproduction of the species.[17] Once propagation of the species becomes the highest ethical goal (a position many followers of Darwin embraced in the late nineteenth and early twentieth centuries), any traits conferring selective advantage for a species become morally desirable and should be preserved, while deleterious biological traits should be extinguished.

The distinction between an ethic and morality is extremely important in analyzing Hitler's worldview and policies, because Hitler's morality sometimes converged with Christian morality (as well as other forms of morality). Because of this, some mistakenly think that Hitler's morality flowed primarily from Christian ethics. Others, who find this explanation too implausible, think that he must have been insincerely pandering to other Germans' moral tastes when taking moral positions. Alternatively, the historian George Mosse argues that Nazi morality emphasized respectability, flowing largely from nineteenth-century bourgeois morality (though he also admits that Nazi morality was ambiguous).[18] All these explanations contain a kernel of truth, but they are only partly true. Yes, Hitler, like just about everyone else in Western culture, never fully escaped the influence of Christian morality and bourgeois ideals. Also, savvy politician that he was, Hitler used any convergence between his morality and that of his fellow Germans to his political advantage.

It should come as no surprise that some of Hitler's morality comported well with traditional Christian morality, since he was, after all, raised in a culture dominated by Catholic Christian morality. I do not deny that some of his moral ideas—for instance, opposition to prostitution or concern for the poor—were shaped in part through the influence of Christianity. However, the similarities often collapse when one asks: How did Hitler justify the view that *x* or *y* is right or wrong (thus getting to the ethic underlying the moral command)? For example, Hitler's opposition to abortion is sometimes portrayed as evidence of his traditional Christian moral views. However, Hitler never appealed to religion, God, or divine revelation to ground his opposition to abortion. Rather he insisted on vigorous enforcement of extant antiabortion laws because he considered German population expansion vital to the improvement of the Aryan race. Also, Hitler did not oppose abortion per se, but only abortion of healthy, Aryan babies. Abortion was permitted—and even encouraged or required—for those who might produce "inferior" offspring or for Jews. The ultimate authority here was not God, the Bible, religious tradition, or any fixed moral code containing the command, "thou shalt not kill." Rather, for Hitler the highest arbiter of morality and political policies was the evolutionary advancement of the human species. In the final analysis, Hitler based his morality on a racist form of evolutionary ethics. Claudia Koonz is right when she argues that the Nazi ethic was a secular replacement and repudiation of traditional Christian ethics.[19]

Aside from Koonz, who properly identifies the centrality of race in the Nazi ethic, few scholars have examined the role of ethics and morality in Hitler's ideology. However, many scholars today agree that Hitler had a coherent worldview that he implemented in radical fashion. Eberhard Jäckel's cogent argument that Hitler embraced a largely consistent worldview that was solidified by the early to mid-1920s has won over many—perhaps most—historians.[20] Even some historians leaning toward structuralism, such as Ian Kershaw, who tend to emphasize impersonal forces rather than the personal role of Hitler, admit that Hitler upheld a consistent ideology that influenced Nazi policy.[21] Indeed, as one examines Hitler's speeches, writings, and conversations with colleagues, the core elements of his worldview remained fixed and unalterable, at least from about 1923 to his death. His overriding goal, shaped by evolutionary ethics, was to improve the human race biologically. In order to achieve this goal, he clung to the following positions throughout his career:

1. An expanding population is biologically beneficial, so the state should promote pro-natalist policies.
2. The biological quality of the German people should be improved through eugenics policies.

3. Germany needs more living space to accommodate the expanding population, and this can only be obtained through military action.
4. Inferior races must give way to superior ones in the struggle for existence, so policies should favor the superior Aryan or Nordic race.
5. Jews are an inferior race, especially in their moral characteristics, so they need to be eliminated—one way or another—from German society.
6. Racial mixture with inferior races must cease, because it leads to biological decline.

Hitler never deviated from these six basic principles during his career, as I will demonstrate throughout this book.

However, though Hitler rigidly adhered to these fixed ideas, he and his regime were not always consistent in their policies. One reason for this is because sometimes the pursuit of one goal conflicted with another. For instance, pro-natalist policies sometimes conflicted with eugenics policies. Militarism might destroy some of the brightest and best of German youth, thus counteracting eugenics policies. It was not always easy for Hitler or his colleagues to judge which policies were more important at any given time, thus leading to conflicts within the regime over which policy to pursue.

Another reason for conflicts in Nazi policy is because it was not always clear what concrete policies would best promote a particular goal. For instance, advocates of eugenics in the early twentieth century debated among themselves what form of marriage was most conducive to improving the biological quality of the German people. While most favored monogamy, a few began advocating polygamy, and yet others believed that freer sexual relations, including adultery and fornication, would help promote greater biological vitality.[22] Nazi officials' views on marriage mirrored this uncertainty. Among Nazi officials there were sometimes sharp disagreements on which policies would best advance the overarching goals set before them. Hitler himself did not always know what policies would best promote his goals, so he often left the details to subordinates. However, he never diverged from the basic goals, and he made sure that they were always being pursued.

Hitler was also very flexible in tactics, which sometimes made the road to Auschwitz very twisted indeed. Especially in the early days of his regime, he was willing to postpone some of his long-range goals if he recognized that they were unobtainable at the moment. However, he never abandoned the long-range goals and constantly looked for the earliest opportunities to implement them. Thus, he did not introduce sweeping anti-Semitic legislation during his first year in office. Rather, he introduced several discriminatory measures in the first year, and then later increased the discrimination step by step. He did not introduce conscription or remilitarize the Rhineland until he deemed it tactically feasible. Some scholars claim that Nazi policies became radicalized in the face of political and military realities during World War II. While this is certainly true, we must keep in mind that the main goals of Hitler and his regime remained unchanged; only the means to achieve the goals became more radical.

Though the core elements of Hitler's ideology were consistent and fixed, he was not a disciplined thinker. He did not always define terms carefully, so this has led to confusion about his ideology (sometimes this served a useful propaganda purpose). Sometimes, especially in his private monologues, he spouted out ideas based on recent events or books he was reading. For the most part, however, in my treatment of Hitler's ethic, I will be focusing on elements of his ideology that were consistent throughout his career. Thus, when I cite various speeches and writings of Hitler to buttress my arguments, I will often choose statements he made at different times of his career. I do this purposely to show that Hitler embraced these concepts throughout his career, not just temporarily. My method of emphasizing the coherent elements of Hitler's ideology may make Hitler seem more consistent than he really was, but I believe it will help us understand his thought and policy better than if we begin picking apart the inconsistencies of his thought (though I will expose these too, at times). Most of the inconsistencies of Hitler's thought related to subordinate policies, not central principles. Hitler's ideology and policies were characterized largely by consistency in the core but inconsistencies in the details.

The biggest inconsistency, however, was between Hitler's personal life and his ideology. Hitler demanded of others what he was unprepared to fulfill himself. He was a hypocrite of colossal proportions. He insisted on the primacy of physical health, but he was rejected for service in the Austrian military because of his poor health. Nazi police arrested vagrants and sent them to concentration camps, even though Hitler had been a vagrant in Vienna before World War I. He idealized labor, calling it the quintessential Aryan trait, but he never held a normal job in his life. Urging every (other) healthy German to marry and procreate as prolifically as possible, he refused to marry (until the last day of his life) and never had any children. His elite SS men had to prove pure Aryan ancestry, but Hitler did not know who his paternal grandfather was, so he was unable to refute rumors that he had Jewish blood (the rumors likely had no basis, but not even Hitler was entirely sure of that at the time). Hitler's brown hair hardly matched the Aryan ideal by which other people were judged. The central elements of Hitler's ideology may have been internally consistent, but he did not seem at all concerned about bringing his own life into harmony with it.

Hitler's ideology was drawn from many different sources, and by no means do I think that evolutionary ethics or social Darwinism were the only culprits responsible for Nazi ideology or practice. Neither do I claim in this book to provide a complete explanation for Nazi ideology. Many other influences shaped Hitler's worldview, including Prussian militarism, German nationalism, Christian anti-Semitism, Gobineau's racism, anti-parliamentarian attitudes, the experience of World War I, and Schopenhauer's philosophy, just to name a few. His thinking was also shaped by many nonrational and noncognitive factors, such as fear, anger, wounded pride, and resentment. However, many of these currents of thought (and feeling)—some predating Darwinism by centuries—were often retooled by social Darwinists in the late nineteenth and early twentieth centuries.

Social Darwinism and evolutionary ethics were so widespread in German and Austrian society and culture by the early twentieth century that it is difficult to identify the precise sources of Hitler's evolutionary ethic. Hitler hardly ever mentioned what books he read or who influenced his thought. This was likely a conscious ploy by him to pose as an original thinker. Hitler's thought, however, was not original at all, and it reflected ideas widespread among radical Pan-German nationalists in Vienna and Munich in the early twentieth century.

We do not know if Hitler ever read Ernst Haeckel, the leading Darwinian biologist in Germany in the late nineteenth and early twentieth centuries. Haeckel consistently argued that ethics should be based on evolution. Further, he was a staunch social Darwinist, arguing from the 1860s on that humans are locked in an ineluctable struggle for existence. He argued that Darwinism proved human inequality, especially racial inequality. He was also the earliest German proponent of eugenics and even euthanasia for the disabled. Since Haeckel's *Riddle of the Universe at the Close of the Nineteenth Century* was a phenomenal bestseller, Hitler could hardly have avoided hearing about Haeckel's ideas. Even if he never read Haeckel, he could easily have imbibed Haeckel's ideas indirectly through the works of many racial thinkers and eugenicists who revered Haeckel.[23]

Social Darwinism was a popular theme in the Viennese press when Hitler lived there as a young man from 1907 to 1913.[24] Though the Viennese racial ideologist Jörg Lanz von Liebenfels was not really *The Man Who Gave Hitler His Ideas*, as Wilfried Daim argued, most historians recognize a measure of parallelism between Lanz von Liebenfels' writings and Hitler's ideology.[25] Lanz von Liebenfels edited a journal dedicated to advancing the cause of the Aryan race. He later claimed that Hitler read his journal, which would not be surprising (though I am not entirely convinced that Lanz von Liebenfels' testimony is reliable, since he also implausibly claimed that Lenin was one of his disciples). He promoted ideas that would feature prominently in Hitler's ideology later, such as social Darwinism, Aryan supremacy, the racial struggle for existence, racial purity, and eugenics.[26] Lanz von Liebenfels' likeminded comrade Guido von List, another Viennese Aryan supremacist, may also have influenced Hitler as a young man. Many other Viennese racial thinkers also could have influenced Hitler, either directly or—more likely—through the press and periodicals. The most prominent of these was the famous racial theorist Houston Stewart Chamberlain. As a young man studying biology under the Darwinian biologist Karl Vogt, Chamberlain enthusiastically embraced Darwinism. Though later he rejected Darwinian evolution as a biological theory, he credited Darwin for hitting on an idea foundational to Chamberlain's racist ideology: the racial struggle for existence.

Probably exerting even greater influence on the development of Hitler's ideology were the racial thinkers in Ludwig Woltmann's social anthropological circle. Woltmann wrote his most important book, *Political Anthropology* (1902), in response to the Krupp Prize Competition, which asked how Darwinian theory should be applied to legislation. He proclaimed in his book, "The same process of natural selection in the struggle for existence rules over the origin, evolution and destruction of human races."[27] He consciously synthesized Gobineau and Darwin, arguing that the Nordic race had reached a higher stage of evolution than other races. The goal of his journal, *Political-Anthropological Review*, was the "application of natural evolutionary theory in the broadest sense of the word to the organic, social, and mental development of peoples (*Völker*)."[28] In the first decade of the twentieth century Woltmann led a circle of committed social Darwinist racists, including the freelance anthropologist Otto Ammon, author of *Natural Selection among Humans* (1893), and the French social Darwinist, Georges Vacher de Lapouge. Ammon was a leading figure in the Pan-German League and helped import social Darwinist ideology into that organization. Not only does the historian Peter Walkenhorst demonstrate that social Darwinist ideology permeated the Pan-German League in the early twentieth century, but he also calls social Darwinism "a central 'background conviction' of Wilhelmine society."[29]

Woltmann's ideas had a profound impact on anthropologists and eugenicists in the early twentieth century, including some of the leading scholars in Weimar and Nazi Germany. Between 1904 and 1907 the psychiatrist Ernst Rüdin wrote very positive reviews of three of Woltmann's books in the leading eugenics journal in Germany.[30] Rüdin later became director of the Kaiser Wilhelm Institute for Psychiatry in Munich and played a leading role in the international eugenics movement before the Nazi period. The Nazis promoted him to lead the German eugenics organization. The anthropologist Eugen Fischer, the founding director of the prestigious Kaiser Wilhelm Institute for Anthropology, Eugenics, and Human Heredity in 1927, confessed that Woltmann—along with Gobineau and Ammon—were important influences on his racial views.[31] After the Nazis came to power, Fischer was named rector of the University of Berlin. He also participated in Nazi government committees and helped implement Nazi racial and eugenics laws. Another leading racial thinker influenced by Woltmann was Ludwig Schemann, the leader of the Gobineau Society. In many of his works, including his monumental history of racial thought published in 1931, Schemann credited his friend Woltmann with being a genius. He also lauded the work of his friends and colleagues Ammon and Lapouge.[32]

We do not know for sure if Hitler ever read any of Woltmann's works. However, a copy of Woltmann's book, *The Germans and the Renaissance in Italy* (1905), at the United States Library of Congress still displays Hitler's bookplate: "Ex Libris Adolf Hitler."[33] Nonetheless, even if he did not personally read Woltmann, the influence of this racial thinker on anthropologists, eugenicists, and the Pan-German nationalist scene was so pervasive that Hitler could hardly have escaped his influence. Woltmann's works were later republished during the Nazi period by the anthropologist Otto Reche, professor at the University of Leipzig and staunch supporter of Nazi racial ideology. When introducing Woltmann's works, Reche claimed that Woltmann was a forerunner of the Nazi worldview. He also confessed that Woltmann's works had profoundly influenced his own thought at the beginning of the twentieth century, making him a disciple.[34]

If Hitler did not already embrace evolutionary ethics before returning to Munich after World War I, he certainly had many opportunities to imbibe these views in right-wing Pan-German nationalist circles in Munich. One of the most prominent Pan-German activists spreading a racialized version of evolutionary ethics was Julius Friedrich Lehmann, who by joining in March 1920 became member number 878 of the Nazi Party.[35] Lehmann used his publishing house to sponsor books, periodicals, and pamphlets promoting scientific racism, eugenics, and anti-Semitism. As a personal friend of Hitler, he sent copies of many of his publications to Hitler, especially ones about racism and eugenics. Lehmann recruited and financially supported Hans F. K. Günther—later one of the leading Nazi racial ideologues—to work on his book, *German Racial Science*. Perhaps more importantly, in 1917 Lehmann began publishing a major journal, *Germany's Renewal*, which featured articles about eugenics, Nordic racism, and racial anti-Semitism. Hitler's library contained an offprint of an article from the 1917 volume by Fritz Lenz, "Race as the Principle of Value: On a Renewal of Ethics," in which Lenz argued that race is the fundamental ethical consideration. Lenz republished this article in 1933, boasting that it "contained all the main features of the National Socialist worldview."[36]

But did Hitler actually read *Germany's Renewal*? Again, we cannot be absolutely certain, but circumstantial evidence suggests that he did. The editor, Erich Kühn, was not only an early member of the Nazi Party, but he was the featured speaker at the first German Workers' Party meeting that Hitler addressed in October 1919.[37] In a circular to Nazi Party members in March 1922, Hitler personally recommended that all party members read *Germany's Renewal*.[38] Not only did Alfred Rosenberg contribute an article to this journal in 1922, but Hitler himself published an article in it in 1924, when the *Völkischer Beobachter* was banned. In January 1924 *Germany's Renewal* carried the famous letter from Houston Stewart Chamberlain to Hitler, where Chamberlain heralded Hitler as Germany's leader. Further, we know that Lehmann gave Hitler many books as they rolled off his presses, so it seems likely that he would have passed on to Hitler copies of *Germany's Renewal*. While Hitler was in Landsberg, Lehmann's son-in-law, Friedrich Weber, was a fellow inmate. Later Hitler often praised Lehmann for his role in helping prepare the ground ideologically for Nazism.[39] With everything we know about the relationship between Hitler and Lehmann, it would be astounding if Hitler did not read *Germany's Renewal*. In any case, Lehmann's influence on Nazi ideology has been noted by many historians. Paul Weindling even called Lehmann's publishing house "a nursery of Nazi racial activists."[40]

In any case, wherever Hitler derived his ideas, his evolutionary ethic was not just one idea among dozens of others tumbling around in his brain. Rather it was a central, foundational idea that provided structure and guidance for many—probably most—of his other ideas and policies. Clearly anti-Semitism, racism, militarism, nationalism, male dominance, and many other Nazi ideas were circulating long before social Darwinism or evolutionary ethics arrived on the scene in the late nineteenth century, so they did not simply derive from evolutionary theory. However, Hitler and other social Darwinists integrated many of these concepts into an overarching worldview that placed the Darwinian struggle for existence (especially between races) at the center of their conceptual universe. Hitler's evolutionary ethic was the guiding principle behind many important policies, including eugenics, population growth, killing the disabled, expansionist warfare, racial struggle, and killing the Jews. As I will show, even his concepts of the People's Community (*Volksgemeinschaft*) and the Leadership Principle (*Führerprinzip*) were integrated into his vision of evolutionary ethics (though they did not derive from biological evolution). My interpretation—that the core of Nazi ideology was evolutionary ethics—shows how all these seemingly disparate programs were linked ideologically.

What about anti-Semitism? Was it not a central guiding principle of Nazi ideology, as some scholars argue?[41] Hitler's anti-Semitism, after all, linked or influenced many important aspects of his ideology, including nationalism, anticommunism, anticapitalism, antidemocracy, and opposition to artistic modernism. However, while anti-Semitism was undoubtedly a prominent factor in Nazi ideology, it cannot explain many important elements of Nazi ideology and practice, such as eugenics, pro-natalism, killing the disabled, militarism, expansionism, or Nazi racial policies aimed at Gypsies, blacks, Slavs, or the "asocial." Hitler's anti-Semitism did not derive from Darwinism, since many elements of the anti-Semitic ideology he embraced predated Darwin by centuries, and even some of the modern elements derived from non-Darwinian sources. Hitler likely imbibed his anti-Semitism from a variety of German thinkers, such as Arthur Schopenhauer, Richard Wagner, and Theodor Fritsch, as well as from anti-Semitic colleagues in Munich, such as Gottfried Feder, Dietrich Eckart, and Julius Friedrich Lehmann. Some elements of his anti-Semitism—especially the stress on a Jewish world conspiracy that included Bolshevism—derived from Russian émigrés, most notably Alfred Rosenberg, who had a profound influence on Hitler in his early political career.[42]

However, though Darwinism did not contribute anything to the origins of anti-Semitism, it did influence the development of anti-Semitic ideology in the late nineteenth and early twentieth century. As many scholars have noted, anti-Semitism was transformed in the late nineteenth century from a religious and social prejudice into a secular racial theory. Some prominent racial anti-Semites in the late nineteenth and early twentieth century, such as Theodor Fritsch and Willibald Hentschel, interpreted their struggle against Jews as a part of an ineluctable Darwinian struggle for existence. Hitler followed this line of thought and integrated his anti-Semitism into a wider vision of evolutionary advancement through the struggle for existence between races and individuals.[43]

Another reason we need to understand the role of evolutionary ethics in Hitler's worldview is because Hitler—like many biologists, anthropologists, and eugenicists of his day—believed that moral characteristics, such as diligence, thrift, and honesty, were biologically innate traits. Thus, evolutionary progress not only brought physical and intellectual advance but also produced moral improvement. Since Hitler considered the Aryan race biologically—and morally—superior to other races, anything that promoted the triumph of the Aryans in the racial struggle for existence was morally good and would produce a more moral world order. The extermination of inferior races would rid the world of the immoral characteristics allegedly rooted in their biological fabric. Hitler thus thought that by killing certain people he could improve the moral stature of humanity. Thus he committed some of the worst atrocities in world history in the name of morality.

CHAPTER ONE

Hitler as Moral Crusader and Liar

Posing as a Moralist

It seems grotesque in retrospect, but Hitler posed as a moral crusader gallantly battling the forces of iniquity, corruption, and even deceit. Many Germans, horrified by the loosening of moral standards in Germany after World War I, were duped by his promises of moral rejuvenation. Hitler's project resonated with many who were disgusted by the rampant hedonism and carnality of Weimar high culture and popular culture. Whether one views Hitler and Nazism as a utopian and technocratic expression of the modernist project, or as an atavistic reaction against modernity, or as some blend of the two ("reactionary modernism" or "conservative revolution"), or as something completely unique, it is clear that Nazism promised a resurrection or awakening of the German people that involved a revival of morality that was in the process of decay and degeneration.[1]

Indeed, Hitler often proclaimed that he stood for morality and decency, preaching the necessity of moral regeneration for the German people. In his very first anti-Semitic writing, a letter written on September 16, 1919, Hitler explained that Germany needed "a rebirth of the moral and intellectual forces of the nation."[2] Three years later he spoke to a Nazi youth organization on "Duty, Loyalty, Obedience," explaining that Germany needed a revival of these three virtues in order to regain its stature.[3] In a 1931 open letter to German Chancellor Heinrich Brüning he stated that the greatest task facing Germany was to gain equality with other nations. This goal could only be achieved, Hitler explained, through "the ethical and moral regeneration of our Volk."[4] After attaining power he explained in a radio speech in October

1933 that the primary goals of the Nazi regime were "restoring order in our own *Volk*, providing work and bread for our starving masses, [and] proclaiming the concepts of honor, loyalty and decency as elements of a moral code of ethics."[5]

These are not isolated examples. Moral regeneration or rebirth was a frequent refrain in Hitler's political sermons. He constantly used morally loaded terminology to portray himself and the Nazi movement as paragons of righteousness. Duty, loyalty, honesty, hard work, orderliness, and cleanliness were virtues he wanted to inculcate in all Germans—one way or another. Even Nazi concentration camps—the epitome of oppression and brutality—carried the aura of moral improvement (though Hitler avoided discussing concentration camps publicly). The Nazi regime always masked the brutality of the camps, publicly portraying them as humane institutions to rehabilitate wayward Germans. According to Himmler, in the concentration camps "there is only one road to freedom. Its milestones are called: Obedience, Diligence, Honesty, Orderliness, Cleanliness, Sobriety, Truthfulness, Self-Sacrifice and Love of the Fatherland."[6] By the mid-1930s many of the inmates of the camps were incarcerated for alcoholism, homosexuality, or for being "asocial," a loosely-defined category including vagrants and prostitutes. Thus they were supposed to be purging German society of its immoral elements. Prisoners entering the gates of Dachau and other concentration camps were greeted by the pious-sounding, but cynical, slogan, "Labor liberates." The Nazified German media regularly depicted the German concentration camps as humane centers where deviants were sent for the inculcation of virtues such as diligence, orderliness, and cleanliness.[7]

Hitler continually depicted his struggle against "Jewish materialism" and Marxism as a principled fight against immorality. He always staked claim to the moral high ground, even when he violated the laws and norms of his society. For example, in 1924, when he was on trial for treason after the Beer Hall Putsch, Hitler freely confessed that he had committed the acts of which he was accused. However, he self-righteously insisted that he was not guilty of any crime, because he was doing good, not evil. He was convicted anyway, but only after gaining considerable popularity for turning the tables on the government by accusing them of being the real criminals. While serving his prison sentence for treason in Landsberg, he wrote an article justifying his role in the putsch attempt. Therein he explained, *"Marxist internationalism will only be broken through a fanatically extreme nationalism of the highest social ethic and morality."*[8] Hitler standing for the highest ethic and morality?

How bizarre this seems in light of later events! At the time, however, many Germans lapped it up.

Hitler's enthusiastic followers looked up to him as the epitome of virtue.[9] Even after the depths of Nazi atrocities had been revealed to the whole world, Hitler's Foreign Minister Joachim Ribbentrop still maintained in his memoirs that Hitler had devoted his entire life to serve the German people. "He lived selflessly, sacrificed his health and, to his last breath, thought of nothing but the future of the nation," Ribbentrop averred.[10] Joseph Goebbels apparently agreed, since he wrote in his diary in October 1939 that Hitler had many virtues, including bravery, a willingness to sacrifice, and contempt for comfort.[11] Even those features of Hitler's personal life that did not agree with his ideology—such as never bearing children—were usually interpreted as altruistic acts of self-denial to benefit the German people.

Many Germans cheered when Hitler promised to clean up the moral depravity of the urban areas and return to the legendary, pristine purity of the village community. His fulminations against prostitution, homosexuality, abortion, and birth control resonated with those Germans who rejected the loose moral standards of the twentieth century. The Nazi press portrayed Hitler as a fervent proponent of "family values." For propaganda purposes Hitler kept his own moral transgressions, especially his relationships with women, carefully hidden from the public. However, his reputation as a teetotaler bolstered his claim to a disciplined, orderly lifestyle. When he purged the SA in 1934, summarily executing many leaders, he castigated them for their homosexuality, as well as for their drunkenness and lavish lifestyles.[12] He thus justified his brutality publicly by claiming he was purging immorality from the midst of the Nazi Party.

Ironically, Hitler's moral crusade included preaching against deceit. He routinely accused his opponents of lying, both about the Nazi Party and about their own policies. He posed as a man of honesty and integrity who would never stoop to breaking promises. On February 10, 1933, in one of his first public speeches to the German people after being named chancellor, he insisted that he would never deceive the German people. In response to those wondering what his political program would look like, he stated that the first point in his program was: "We do not want to lie and we do not want to con." That is why, he continued, he never made cheap promises.[13]

Hitler hoped to gain the German people's trust by posing as a principled truth-teller. Lying only works politically if the public believes the lie, and Hitler knew this. However, surprisingly, honesty actually did play a role in Hitler's worldview. He considered honesty an important trait characterizing the noble Aryan race. Even more frequently he fulminated against the Jews for their deception. They were not like the Aryans, he claimed, since Jews cunningly engaged in deceit and trickery at every opportunity, often duping the honest but sometimes naïve Aryan. In *Mein Kampf* Hitler contrasted the honest Aryans and lying Jews, stating, "But the means with which he [the Jew] seeks to break such reckless but upright souls [Aryans] is not honest warfare, but lies and slander."[14] In another passage he accused the Jews of combining "bestial cruelty and an inconceivable gift for lying" in their inherent racial character. Multiple times he followed Schopenhauer in calling the Jews "the great masters in lying." All of Jewish existence is based on a lie, according to Hitler, since they pretend to be a religious community when they are really a race.[15]

Hitler also insisted that his own worldview represented "eternal truth" in a struggle against the wiles of other worldviews, especially Marxism.[16] Many of his followers considered him, unlike other politicians, a resolute defender of the truth. In discussing the role of his own speeches in swaying the crowds, Hitler indicated that he was inculcating

Figure 1.1   Nazi poster: "Death to the Lies" with "high finance" on the serpent's back and "Marxism" on its belly

in the masses a worldview and philosophy that was true and valid. He claimed he was replacing a false Marxist internationalism with a racial nationalism that corresponded with the facts of history, science, and current events. When discussing one of his early lectures on the Versailles Treaty in Munich, he asserted, "Again a great lie had been torn out of the hearts and brains of a crowd numbering thousands, and a truth implanted in its place."[17] While other politicians were hypocritical, self-serving, or pandered to the rich or the Jews (or both), Hitler claimed to be unswervingly embracing and advancing the cause of the truth. He fostered this image in *Mein Kampf* by stressing his unwavering commitment to an unchanging worldview. Later, he also depicted World War II as a battle of truth against the mendacious propaganda of the British and Jews. In October 1941 he told a crowd in Berlin that "a struggle between the truth and the lie has been taking place. As always, this struggle will also end victoriously for the truth."[18]

Hitler as Liar

By now you are probably shaking your head, thinking I am confusing mendacious propaganda with heartfelt conviction. How could I actually believe that Hitler was serious when he piously proclaimed that he stood for the "highest social ethic and morality"? After all, wasn't Hitler a consummate Machiavellian politician willing to use any means necessary to achieve and maintain power? Wasn't he more concerned with the effects of his words on the masses than he was with truth? I am under no illusions, for I recognize that Hitler had no moral qualms about lying. If lies were effective in helping him attain his goals, then he lied with gusto. Claiming to be offended by those impugning his honesty was nothing but another pose calculated to deflect the attacks by his enemies.

Not very often did Hitler divulge his thoughts about the propriety of lying to achieve his goals. After all, lies are only effective if the other party assumes one is not lying, so Hitler had to maintain his image of truthfulness, inasmuch as that was possible. However, at times Hitler made known his lack of concern for truthfulness. In one passage of *Mein Kampf* Hitler criticized the German regime for their war propaganda during World War I, which he considered far inferior to that of Britain. "Propaganda in the War," Hitler asseverated,

> was a means to an end, and the end was the struggle for the existence of the German people; consequently, propaganda could only be considered in accordance with the principles that were valid

for this struggle. In this case the most cruel [sic] weapons were humane if they brought about a quicker victory; and only those methods were beautiful which helped the nation safeguard the dignity of its freedom. This was the only possible attitude toward war propaganda in a life-and-death struggle like ours.[19]

This implies that any means, including lying, is allowed in the struggle for existence. For Hitler the stakes had been high in this particular case, leading to Germany's loss in World War I. In April 1939 Hitler reiterated his view that Germany had been defeated by deceit, not by military might, in World War I.[20]

Randy Bytwerk is correct when he asserts, "Hitler did not advocate lying as a general principle, though he saw it as a sometimes necessary tool."[21] In a couple of passages of *Mein Kampf* Hitler made explicit his view that untrue statements are permissible in propaganda. When discussing war propaganda, he stated that "it would have been correct to load every bit of the blame on the shoulders of the enemy [in World War I], even if this had not really corresponded to the true facts, as it actually did."[22] Facts apparently should never get in the way of propaganda, which can only be judged according to its effects. Any propaganda was good if it met his larger goals. When discussing whether or not the Nazi Party should consider revising its Twenty-Five Point Program, Hitler further manifested disregard for truth. Even if we discovered that some points in the program "should not entirely correspond to reality," he wrote, the entire program must remain unaltered. The Nazi movement—and Hitler as its leader—must never be seen as fallible. Also, putting the party program up for discussion would sap strength from the movement, thus distracting it from pursuing its primary goals.[23]

How Can We Know What Hitler Believed?

This penchant for lying presents the historian with methodological difficulties, especially when trying to construct Hitler's "real" position on anything. How can we believe anything Hitler says? How can we separate his propaganda from his real worldview? While acknowledging the perils of trying to fathom the thoughts of a man who purposely remained aloof and elusive and who deliberately lied to conceal or distort his own position whenever he wanted, I do not think that we must simply throw up our hands in despair. I am not convinced that Hitler's worldview is completely inscrutable. Good historians should always proceed with a healthy dose of skepticism when examining documents, so by applying the same critical methods used to interpret other historical documents, we might be able to gain some knowledge about Hitler's ideas. Yes, we must be cautious in proceeding, but if we are careful, I think we can advance.

I propose several ways to determine what Hitler really believed. First, we must compare Hitler's statements over his entire career. Was he consistent over time in his pronouncements about a specific idea or policy, or not? Why or why not? If he was inconsistent, did he change his views, or was he lying (in one or both cases)? Second, we need to consider his intended audience. His public speeches often contain statements at odds with the positions he divulged in private conferences and conversations. Many times it is obvious that he was lying to the German public or the international community, simply telling them what they wanted to hear, rather than revealing his own convictions. In private speeches to the party faithful, however, he often was more frank about his worldview. Third, we need to compare his statements with his actions. Did Hitler implement policies consistent with the ideas he expressed in his speeches or in the party platform, or not? Finally, we always need to ask ourselves if a particular position Hitler took publicly would bring him political advantage. If so, we should be a little more skeptical about it.

Let us apply this method to an area where Hitler's lies are, in retrospect, glaringly obvious. After gaining power in 1933, Hitler consistently portrayed himself as a man of peace. He repeatedly assured the world of his peaceful intentions and signed several nonaggression pacts, promising never to attack his neighbors. Not only did he profusely spout out promises of peace, but he continually insisted that he was a man of honor and would never break his word. He admitted that he might break treaties that had been forced upon Germany before he came to power (meaning, of course, the Versailles Treaty), but his word was his bond. As one example among many, in February 1935 Hitler delivered a speech in Munich, presenting himself and Germany as peace-loving and respectful of the rights of other nations. After explaining that he only wanted equality with other nations, a frequent theme in his speeches in the mid-1930s, he stated that "the world can also rest assured that, when we do sign something, we adhere to it. Whatever we believe we cannot adhere to, on principles of honor or ability, we will never sign. Whatever we have once signed we will blindly and faithfully fulfill."[24] Hitler was still repeating his nonaggression mantra in late January 1939, less than two months before forcing Czechoslovakia to capitulate and just seven months prior to attacking Poland. He even hypocritically accused his critics of lying when they charged him with harboring aggressive intentions.[25] Immediately after Germany took over Czechoslovakia and divided it up in March 1939, General Brauchitsch learned that Hitler was not sure how long he would honor the newly signed treaty with Slovakia.[26] Even after violating his guarantees to Czechoslovakia, Hitler signed a nonaggression pact with Denmark in late May 1939, but he violated it less than a year later. Hitler continued giving false assurances to various countries—including his archenemy, the Soviet Union—as long as he thought it would benefit the German war effort.

We all know today that Hitler was lying about his peaceful intentions, but how do we know that? First, before coming to power, Hitler often stressed the need to expand Germany's borders, by force if necessary. This was an integral part of his worldview (see chapter 8). Second, we know that it served Hitler's political purposes to lie to the international community about his militarist and expansionist intentions, because he wanted to allay their fears while abolishing the Versailles Treaty and rearming Germany. He could not afford to arouse any vigorous opposition from France and Britain before he had fully armed. Third, Hitler privately spoke about the need for expansion by military means, even while denying it publicly. Fourth, he repeatedly prepared for wars of aggression even while stating publicly that he had no designs on any other nation's territory. Every time Hitler made bold foreign policy moves—remilitarizing the Rhineland, annexing Austria, taking over the Sudetenland from Czechoslovakia, occupying Czechoslovakia, and so on—he always claimed he had no further demands and would not attack anyone. However, behind the scenes he was already preparing for his next aggressive act. Fifth, after launching warfare against his neighbors, he often described his wars of expansion as natural and necessary.

We also know that Hitler was lying because we now have access to private speeches and conversations during which Hitler admitted that he lied about his peaceful intentions. In a revealing private speech to representatives of the Nazified German press in Munich in November 1938, Hitler told them—less than ten months before he began World War II by attacking Poland—that one of their main tasks in the upcoming year was to prepare the German people for war. He explained that he had been forced by circumstances to preach peace, because this was the only way to rearm Germany. His peace propaganda was obviously intended primarily for international consumption, and indeed many foreign powers gobbled it up, hook, line and sinker. However, Hitler noted that his peace propaganda had produced a negative side effect, because some Germans were taking it seriously![27] Hitler's peace propaganda apparently succeeded too well, for when the campaign against Poland opened on September 1, 1939, most Germans were sullen. Hitler was disgusted by their lack of enthusiasm for war.

Another time that Hitler revealed the duplicity of his public propaganda about peace was in a private speech to Nazi Party leaders in April 1937. He stated:

> We all know that there are some things about which we should never speak.... We know certainly, that we are building our army up, in order to keep the peace. And we are running the Four Year Plan, in order, we say, to be able to exist economically. Only *thus* can we speak of these matters. Each of us knows that. Other thoughts will never be uttered, and that is true in very many areas. This must be an iron principle. Each one [of us] can look the other in the eye, and he can from the eyes perceive, that the other thinks exactly the same way that he thinks, and knows exactly the same as he also knows.[28]

Here Hitler was commanding the party faithful to keep their mouths shut about real Nazi intentions concerning remilitarization and the Four Year Plan, which aimed at putting the German economy on a war footing. Essentially he condoned, encouraged, and even required deception. He also told them that this principle of silence (and, by implication, lying) applied to "very many areas," not just this particular case. After giving his fellow Nazis the green light to lie to others, however, in the same speech he urged them to be fanatically loyal to each other, eschewing deceit and trickery.[29]

Hitler admitted to his military leaders that treaties were not binding, thus confessing that lying was official policy in diplomacy. In November 1939, while discussing the Nazi-Soviet Nonaggression Pact, he told them, "Treaties, however, are kept as long as they serve a purpose."[30] This should have been obvious by that time, since he had already violated the Munich Agreement, his earlier Nonaggression Pact with Poland, as well as other treaties he had negotiated.

Hitler violated so many treaties and agreements after 1938 that today Chamberlain's appeasement policy seems exceedingly gullible. How could anyone have trusted Hitler to keep his word? In retrospect, it was naïve, but until 1938 Hitler kept his international agreements, at least as far as anyone could tell. The invasion of Czechoslovakia in March 1939 was the first clear violation of Hitler's international promises, since he had guaranteed Czechoslovakia's integrity in the Munich Agreement just a half year earlier. However, even after brazenly breaking his promise, Hitler still tried to smooth things over with the British. He did not want them intervening when he staged his next war of expansion against Poland. Shortly before that military campaign began, when a diplomat told him that England did not trust him, Hitler retorted, "Idiots, have I ever told a lie in my life?"[31] Hitler's pose as a truth-teller did not impress the British this time, since he had already exhausted their gullibility.

With his propensity for lying, was Hitler simply hypocritical every time he exalted honesty and encouraged truthfulness? In one sense, of course he was, for he consciously lied, even in proclaiming his truthfulness. However, there is another way to construe (but not condone) Hitler's penchant for lying. As I will prove in detail in the rest of this book, Hitler was committed to two moral principles that justified his lying, at least in his own mind. First, he believed that moral principles, such as honesty, are always subservient to a higher ethical principle: the evolutionary progress of humanity. Second, he was convinced that morality was only applicable within one's racial community.

Concerning the first point, Hitler rarely (if ever) explicitly used the language of evolutionary progress to justify his lying. However, as we shall see in greater detail later, he did quite frequently make it clear that the highest ethical principle in his worldview was evolutionary progress (see chapter 2). He also explained that he considered lying justified if it brought success to Germany in its quest for expansion of its living space (*Lebensraum*), and expansion of living space was an idea built on social Darwinist principles. In a speech to his generals in August 1939, shortly before attacking Poland, he informed them that he would provide a pretext for the war, and he did not care if it was credible (and it was not). Victors, he explained, are not asked later if they had told the truth. Rather, "with the origin and conduct of the war only victory is relevant, not righteousness (*Recht*)." He then continued by encouraging them to "close your hearts to pity" and "act brutally," so that the German people would "obtain what is their right. Their existence must be secured. The stronger is right."[32] This speech clearly revealed Hitler's philosophy of the ends justifying the means. He had already articulated this point long before coming to power, when he explained the purpose of diplomacy in *Mein Kampf*: "*Diplomacy must see to it that a people does not heroically perish, but is practically preserved. Every road that leads to this is then expedient, and not taking it must be characterized as criminal neglect of duty.*"[33] If "every road" leading to the success of his nation was permissible, surely this includes deceit and duplicity. Hitler could, and I believe he did, consider honesty a valid moral principle, but it took second place to a higher principle: the preservation and expansion of the Aryan race.

Lying could also be morally justified in some circumstances, according to Hitler's line of thought, because not all humanity was included in the moral community. Moral behavior was only required within one's *Volk* or race, terms Hitler used interchangeably. According to Dietrich Eckart, whom Hitler called his mentor, Hitler once asserted that because of Jewish influence, Luther mistranslated the biblical command, "Love your neighbor as yourself." Rather, it should be translated, according to Hitler, as "Love your racial comrade (*Volksgenosse*) as yourself."[34] Though Eckart's conversation with Hitler is likely somewhat fictionalized, both Eckart and Hitler shared this notion of a racial ethic.[35] According to their view, one's moral obligations only extend to members of the Aryan race, not to those of other races.

But didn't Hitler also lie—often and flagrantly—to his fellow Germans? How could he justify that, if lying to one's racial comrades is immoral? Here, it seems that the first principle—promoting evolutionary progress—trumped any consideration for the truth. Hitler thought that the triumph of the Nazi Party in Germany, and then the triumph of Germany against its neighbors, would benefit the highest race, the Aryans, and thus lead to a higher level of humanity. If lying was necessary to achieve this, so be it.

Lying to Jews, whom Hitler clearly did not include in the moral community, was clearly permissible, he thought. Though he often described the Jews as perfidious deceivers, he considered it perfectly acceptable to fight fire with fire. In a 1922 meeting in Munich, Hitler said he did not understand the statement "that one should not proceed with violence against the Jews. We will fight the Jews with the same means that they use against us."[36] If violence can be met with violence, presumably lying can be met with lying. In 1923 Hitler called his fellow Germans to avenge the Jews for their deception: "Our feeling of righteousness demands that this deception of an entire people [*Volk*] be atoned! We will not stop sharpening consciences and arousing emotions! And the day will come that we will destroy the deceivers!"[37] Thus, Hitler's campaign against the allegedly deceitful Jews was framed as a moral crusade.

Hitler's dual position that the end justifies the means and that the racial community defines morality was clearly stated in a 1923 speech, when he remarked:

> But we National Socialists stand here [on the Jewish Question] at an extreme position. We know only one people (*Volk*), for whom we fight, and that is our own. Perhaps we are inhumane! But if we save Germany, we have accomplished the greatest deed in the world. Perhaps we perpetrate injustice! But if we save Germany, we have abolished the greatest injustice of the world. Perhaps we are immoral! But if our people (*Volk*) is saved, we have paved the way again for morality.[38]

Thus he enjoined inhumaneness, injustice, and immorality toward those branded as enemies of the German people. However, he also asserted that this immorality would serve a higher moral purpose. He was not promoting nihilism, since he really did have an end in mind that these immoral means were supposed to serve. This end—indeed the highest goal for Hitler—was evolutionary progress.

Thus Hitler was not just exercising power for power's sake. He continually insisted that he was promoting a consistent worldview. He even told the party faithful at the 1934 Nuremberg Party Rally that "National Socialism is a worldview." Earlier in the same speech he explained that revolutions should not merely destroy, but they need to create new conditions for life. "Woe," he continued, "if the act of destruction does not result in the service of a better and thus higher idea, but rather only exclusively obeys the nihilistic urges of destruction."[39] Is this just another one of Hitler's big lies, another example of his masterful use of propaganda? Though some people think so, many historians—perhaps most—have come to agree with Eberhard Jäckel's position that Hitler really did have a coherent worldview (though historians differ over the significance of his worldview). Hitler really was pursuing what he considered a "higher idea." If we carefully sift through his public and private statements and compare these with his policies, there is remarkable consistency in his goals, even if wavering and inconsistency sometimes characterized his tactics and timing. Hitler sometimes dithered, sometimes waffled, and sometimes did not know how to proceed. However, his goals were always fixed and unshakable.

Because of his propensity for lying, we must be wary of everything Hitler said. However, those who see everything he said or wrote as "merely propaganda" miss an important point. Propaganda has two purposes that are sometimes harmonious, but often are in conflict: (1) to gain political support; and (2) to convince people of one's own position. Lying may at times be an effective way to accomplish the first goal, but it backfires if one wants to bring people to adopt one's beliefs and convictions. If one examines how the Nazis used their propaganda in the educational system once they came to power, it is apparent that they were interested in molding the minds of the German youth to embrace a coherent Nazi worldview. That worldview revolved around evolutionary ethics, as I will explain in the following chapters.

CHAPTER TWO

The Cult of Evolutionary Progress

The Centrality of the Struggle for Existence in Hitler's Worldview

Kampf, meaning struggle or battle, was one of Hitler's favorite words. In the title of his only published book, *Mein Kampf*, it referred to his own personal and political struggles, as is evident from the original title Hitler gave it: "Four-and-a-half Years of Struggle against Lies, Stupidity, and Cowardice."[1] However, for Hitler struggle meant far more than his own conflicts. One of the main themes of *Mein Kampf*, as well as his unpublished *Second Book* and many of his speeches, is the centrality of struggle, including struggle between individuals within society, struggle between nations, and racial struggle. He argued in *Mein Kampf* that a human "must never fall into the lunacy of believing that he has really risen to be lord and master of Nature," but must "understand the fundamental necessity of Nature's rule, and realize how much his existence is subjected to these laws of eternal fight and upward struggle." Nature's laws, especially the law of struggle, are supreme, Hitler asserted, and "there can be no special laws for man. For him, too, the eternal principles of this ultimate wisdom hold sway. He can try to comprehend them; but escape them, never." Instead of bucking against the "ultimate wisdom" of nature and its laws, humans should submit to them. Opting out of the struggle for existence was simply not an option.[2]

In his speeches and writings, Hitler continually stressed the necessity and importance of vigorous struggle in the lives of individuals and the state. Struggle meant more to Hitler than terror tactics to attain power

Figure 2.1 "Life Requires struggle"

or, once he was in power, state-directed violence to suppress opposition. Though using the term struggle in various ways in his speeches and writings, he often portrayed it as a universal law of nature, from which there is no escape. He regularly quoted the famous statement by the Greek philosopher Heraclitus that struggle is the father of all things (though once Hitler ascribed this saying to the famous German military philosopher Clausewitz). He frequently referred to the "law of struggle," the "eternal struggle," and "inescapable struggle." He also liked Darwin's two phrases, "struggle for existence" and "struggle for life," which appear repeatedly in *Mein Kampf* and his speeches, along with another favorite word derived from Darwinian thought: "natural selection," which he often abbreviated simply as "selection."

For Hitler the Darwinian struggle for existence was more than just a phrase to justify violent competition. It lay at the heart of his worldview, coloring almost every dimension of his ideology and policy. In a March 1927 speech he explained the importance of the struggle for existence, both for the individual and for the *Volk*:

> Politics is the striving and struggle of a *Volk* for its daily bread and its existence in the world, just as the individual devotes its entire life to the struggle for existence, for its daily bread. And then comes a second matter, caring for future survival, caring for the child. It is the struggle for the moment and the struggle for posterity. And all thinking and all planning serve in the deepest sense this struggle for the preservation of life.[3]

As this statement and many other similar ones by Hitler made clear, all his policies and plans served one purpose: the success of his *Volk* in the struggle for existence.

He often explained the struggle for existence in ways quite similar to Darwinian biologists and Darwinian-inspired social thinkers in the early twentieth century, though many Darwinists in his own day (and certainly later) were horrified by some of the ways he applied Darwinism to politics. Though Hitler was not particularly astute scientifically, his general explanations of evolutionary competition driven by population imbalances did accurately reflect the scientific thought of his day. His penchant for violent expressions of the struggle for existence, however, was not a necessary corollary of Darwinian theory, though it did not contradict Darwinism.

In *Mein Kampf* and in many of his speeches, Hitler explained that the struggle for existence among organisms is caused by the tendency for populations to expand faster than their food supplies. This was a central idea of Darwin, derived from Thomas Robert Malthus's famous population essay. Malthus had argued that the reproductive rates of organisms—including humans—leads necessarily to competition for scarce resources. However, whereas Malthus portrayed this competition as the cause of misery, poverty, war, and famine, Darwin put an optimistic spin on it. According to Darwin, all the misery had an ultimately positive effect, for it ultimately produced evolutionary progress. In *The Origin of Species* he concluded, "Thus, from the war of nature, from famine and death, the most exalted object which we are capable of conceiving, namely, the production of the higher animals, directly follows."[4]

Social Darwinists in the late nineteenth century—including Darwin himself—believed that the struggle for existence resulted in evolutionary progress for humans, too. In *The Descent of Man* Darwin applied the concept of the struggle for existence to humanity, stating,

> Natural selection follows from the struggle for existence; and this from a rapid rate of increase. It is impossible not bitterly to regret, but whether wisely is another question, the rate at which man tends to increase; for this leads in barbarous tribes to infanticide and many other evils, and in civilised nations to abject poverty, celibacy, and to the late marriages of the prudent. But as man suffers from the same physical evils with the lower animals, he has no right to expect an immunity from the evils consequent on the struggle for existence. Had he not been subjected to natural selection, assuredly he would never have attained to the rank of manhood.... it may well be doubted whether the most favorable [circumstances] would have sufficed [to produce human evolution], had not the rate of increase been rapid, and the consequent struggle for existence severe to an extreme degree.[5]

Darwin never advocated using brutality in the human struggle for existence, and he certainly never called for purposeful killing of the "unfit"—as some more radical Darwinists did later. However, he did recognize that the struggle could be severe and produces "evils," which he considered necessary for further progress.

Some scholars wrongly claim that even though Darwin recognized the force of human competition in producing the human species as it is today, he did not advocate any policy on this basis. This is simply untrue. While not a prominent feature in Darwin's work, he did occasionally state his own position about how his biological ideas should be applied to humanity. In the conclusion to *Descent of Man*, for example, he stated,

> Man, like every other animal, has no doubt advanced to his present high condition through a struggle for existence consequent on his rapid multiplication; and if he is to advance still higher he must remain subject to a severe struggle. Otherwise he would soon sink into indolence, and the more highly-gifted men would not be more successful in the battle of life than the less gifted. Hence our natural rate of increase, though leading to many and obvious evils, must not be greatly diminished by any means. There should be open competition for all men; and the most able should not be prevented by laws or customs from succeeding best and rearing the largest number of offspring.[6]

Darwin would no doubt have been horrified by the way Hitler applied his theory to humanity, but nonetheless, this passage contains several ideas that would later be central to Hitler's ideology. Darwin not only insisted that humans have attained their present rank via "rapid multiplication" causing a struggle for existence, but he also claimed that this "battle of life" is necessary to continue evolutionary progress. Based on this analysis, he then spelled out two implications of his theory for public policy, both of which would be central to Hitler's social Darwinist ideology: (1) maintaining high reproductive levels; and (2) maintaining human competition. By using the terms "must not be," "should be," and "should not be," Darwin crossed the is-ought gap and promoted morality and legislation based on his biological theory.[7]

Hitler may never have read Darwin, but these Darwinian ideas were widespread in Germany in the late nineteenth and early twentieth centuries. Many German scientists and social thinkers exulted in the struggle for existence as a beneficial force in human history. Ernst Haeckel not only claimed that the human struggle for existence produced progress, but he also believed that Darwinism had demolished "anthropocentrism," the view that humans are special or sacred. The prominent ethnologist Friedrich von Hellwald applied Darwinism to human history in *History of Culture* (1875), arguing that "the right of the stronger is a natural law." In an article on the human struggle for existence, Hellwald noted that evolutionary progress would occur as "fitter" humans "stride across the corpses of the vanquished; that is natural law."[8] By the early twentieth century the idea that the human struggle for existence was beneficial was commonplace among German biologists, anthropologists, eugenicists, racial theorists, and other social thinkers. To provide one example among thousands, the leader of the Pan-German League, Ernst Hasse, stated in a 1906 book that the world belongs to the strong and mighty, while the weak disappear: "The struggle for existence is a natural, rational, and [morally] justified process."[9] These ideas circulated widely in the circles Hitler frequented.

Hitler embraced the social Darwinist idea of the struggle for existence as a positive force, bringing progress and improvement to biological organisms, including the human species. He promoted this idea in dozens of his public and private speeches, as well as in *Mein Kampf* and in his *Second Book*. Though he never used the term "Darwinism," he often used the term "evolution" (*Entwicklung*) and even "higher evolution" (*Höherentwicklung*) in his discussions of biological change. Unfortunately the standard English translation of *Mein Kampf* consistently translates the term *Entwicklung* as "development," even in contexts where it clearly refers to biological evolution. Although "development" is an accurate translation of *Entwicklung* in some contexts, *Entwicklung* was the standard term German biologists used for "evolution."[10] The translator of Hitler's *Second Book* often renders "*Entwicklung*" as "evolution."

Hitler consistently and persistently underlined the importance of struggle for human evolution. In a typical statement in a 1927 speech he called the "eternal struggle...the precondition for the evolution of all humanity."[11] Struggle played such a central role in his world-view that in his *Second Book* he devoted the first chapter to "War and Peace in the Struggle for Life." The second chapter, "Struggle, not the Economy, Secures Life," continued this theme, as did chapter three on "Race, Struggle, and Power." He began the first chapter by explaining: "Politics is history in the making. History itself represents the progression of a people's (*Volk*'s) struggle for life." He then explained that this struggle for life is caused by the twin human instincts of self-preservation and reproduction. However, while the reproductive instinct is unlimited, space is limited. Thus, "in the limitation of this living space (*Lebensraum*) lies the compulsion for the struggle for life, and the struggle for life, in turn, contains the precondition for evolution."[12] Hitler had already explained this same point repeatedly in *Mein Kampf*, where he depicted the struggle for existence between humans as a positive force, because it got rid of the weak and sick, preserving only the healthy, and thus producing "higher evolution" (*Höherentwicklung*).[13]

DIE GANZE NATUR IST EIN GEWALTIGES RINGEN
ZWISCHEN KRAFT UND SCHWÄCHE,
EIN EWIGER SIEG DES STARKEN ÜBER DEN SCHWACHEN.

ADOLF HITLER

Figure 2.2 "All of nature is a powerful struggle between power and weakness, an eternal victory of the strong over the weak." Adolf Hitler

Indeed throughout his career he repeatedly invoked struggle as the central, driving principle in the cosmos, and whenever he overtly discussed his own world view, struggle featured as one of the foundational tenets. In a July 1927 speech on the "Essence and Goal of National Socialism," Hitler asserted that struggle produces everything good, because it selects the best. He then stated: *"Imperialism is the struggle for existence of the nation,* ... making it possible to feed itself and reproduce." He quickly added that Nazism upheld a "world view of the natural powers of evolution."[14] Hitler thus presented biological struggle in the evolutionary process as a central tenet of Nazism.

In a speech the following month on "What is National Socialism?" he again stressed the importance of struggle in the Nazi worldview. He opened the speech by explaining—as he did earlier in *Mein Kampf* and the *Second Book*—that the two main forces ruling human life are hunger and love. "Both of these," he asserted, "are grounded in the instinct for self-preservation and the instinct for reproduction." In order to fulfill these instincts, all organisms, including humans, must struggle, because there is insufficient space available for everyone to keep reproducing. "We confess that imperialism is the most natural thing that there can be," Hitler maintained, "because every father, who begets a child, and every mother, who bears a child, and desires that this child lives, is thereby imperialistic, if they want the wider community of their people (*Volk*) to receive bread."[15] Hitler was clearly obsessed with the Malthusian population principle that Darwin had integrated into his theory, and he used it to justify expansionism.

Hitler was still repeating these themes in secret speeches to military leaders during World War II. In a secret speech to 10,000 new military officers on May 30, 1942—while German armies were locked in a bitter war on the Eastern Front and while German forces were simultaneously exterminating Jews and others—he tackled the question, "Was the Second World War Avoidable for Germany?" His answer was that, no, the war was inevitable, since we are constrained by natural laws, including the struggle for existence. His speech was laced with Darwinian terminology, such as evolution, struggle, and selection. His opening remarks divulged his social Darwinist outlook:

A deeply serious principle of a great military philosopher states, that *struggle* and thus war is *the father of all things.* Whoever casts even a glance at nature as it is, will find this principle confirmed as *valid for all organisms and for all happenings* not only *on this earth,* but even far beyond it. The entire universe appears to be ruled only by this one idea, that *eternal selection* takes place, *in which the stronger in the end preserves its life* and the right to life, and the weaker falls.[16]

He then informed these officers that the struggle over territory pits one *Volk* against another and leads to an "eternal selection, to the selection of *the best and hardest.* Thus we see in this struggle an element of the formation of every living thing and even of life itself." By eliminating the weaker and strengthening the stronger, this struggle, Hitler continued, produces "evolutionary progress" *(Vorwärtsentwicklung).*[17]

For Hitler the struggle for existence took on religious dimensions. In a speech on November 1943 in Munich, Hitler countered criticisms coming from religious quarters. He assured his audience that he also was religious, indeed "profoundly religious on the inside." Then he equated the judgment of Providence with the struggle for existence. Providence, he explained, weighed humanity by natural means: selection of the stronger.[18] Selection in the evolutionary process, then, was God's way of working, or so Hitler thought. In another speech about a year later he insisted similarly that Providence only helped those who would fight to win in the struggle. He stated,

> Insofar as the Almighty opened our eyes in order to grant us insight into the laws of his rule, in accordance with the limited capabilities of us human beings, we recognize the incorruptible justice which gives life as a final reward only to those who are willing and ready to give a life for a life. Whether man agrees to or rejects this harsh law makes absolutely no difference. Man cannot change it; whoever tries to withdraw from this struggle for life does not erase the law but only [eliminates] the basis of his own existence.[19]

Hitler then continued by explaining his view that the struggle in nature is ultimately over space, and those who are biologically weak will be restricted in their living space, while the stronger will occupy as much space as they can. He claimed that the stronger groups of people who take over more space are simply following a "command of Providence." In a proclamation of February 1945 Hitler linked success in the struggle for existence to Providence again: "Providence does not show mercy toward the weak. Instead, it only recognizes the right to live for the healthy and strong!"[20]

Some might object that these religious statements of Hitler are merely propaganda. This would not be startling, since in public his statements about religion were consistently positive, while his private utterances to close colleagues often manifested disdain for organized religion. Even if the above religious statements are insincere, it would only strengthen my point that evolutionary progress was of paramount value for Hitler. However, what if Richard Steigmann-Gall is correct about Hitler's affinities to theism and Christianity?[21] Would this pose any problem for my interpretation of Hitler's ethic? Not at all. As the above statements by Hitler clarify, even if he did uphold a theistic position, he thought God ruled primarily through natural laws, including biological evolution. If he was a theist, he was a theistic evolutionist. Even Steigmann-Gall concedes that the form of Christianity most appealing to Nazis was liberal Protestantism. Most liberal Protestants embraced Darwinism with alacrity.

Further, Hitler saw evolutionary ethics as the expression of the will of God. In a 1942 speech in which he discussed the natural law of struggle at length, he equated the laws of nature and the will of Providence. He stated that in order for one organism to live, another must die. If someone would try to counter this natural process, then "nature, Providence, do not ask for his interpretation or his desires, it only knows one law: 'Man, struggle, secure your place in life, then you will live!' Or refuse to struggle, lose your place in life, then you will die and another will replace you." In this same passage Hitler portrayed struggle as a beneficent force, despite the death and misery it causes. There is no better principle imaginable, he argued, than "the principle of the eternal selection of the better over the weaker." Indeed, he explicitly called this principle the "will of Providence."[22] (While it goes far beyond the scope of this work to explain Hitler's religious views in detail, I intend to treat this subject in a later book, where I intend to demonstrate that Hitler was neither an atheist nor a Christian.)

Hitler's philosophy of struggle was not mere propaganda for public consumption. He often stated similar views in private meetings and conversations. Hitler's personal press chief Otto Dietrich recalled that Hitler perpetually talked about struggle, both in public and private. According to Dietrich,

Among Hitler's own justifications for his actions was his primitive philosophy of nature. Both in public speeches and private conversations he would repeatedly refer to this philosophy, his purpose being to convince his listeners that this philosophy represented the final truth about life. He took such principles as the struggle for existence, the survival of the fittest and strongest, for the law of nature and considered them a "higher imperative" which should also rule in the community life of men. It followed for him that might was right, that his own violent methods were therefore absolutely in keeping with the laws of nature.[23]

Thus, if Dietrich is right, Hitler based his morality on the laws of nature, especially those laws propounded by Darwin. Dietrich's perceptive analysis based on his own experiences with Hitler jibes well with my own retrospective conclusions: Hitler exalted the evolutionary process above any other moral considerations. Dietrich calls this his "higher imperative."

One of Hitler's secretaries, Traudl Junge, confirms Dietrich's conclusions. After mentioning that Hitler often led interesting discussions with his entourage about the church and human evolution, she noted that Hitler had contempt for the church. Rather, "his religion was the laws of nature," according to Junge. She then explained that the law of nature Hitler invoked most often was the law of struggle, which humans could never escape, because we are "children of nature." These laws had brought about evolutionary progress, but only by eliminating the weak and those unfit to live. Hitler would also criticize the churches for taking it upon themselves to protect the lives of the weak, the "inferior," and those unfit for life.[24]

Another close colleague of Hitler's, Otto Wagener, who served as chief of staff of the Nazi SA (stormtroopers) from 1929 to 1933, likewise remembered Hitler as radically committed to a worldview that emphasized the necessity of struggle among humans, which would lead to the triumph of the healthier and better. Wagener, who remained committed to the socialist agenda of the National Socialists, explained that even though Hitler shared his commitment to socialism, Hitler "had conflicted feelings," because he simultaneously upheld the necessity of the struggle for existence and applied this principle even to economics. Wagener's claim that Hitler was conflicted probably says more about Wagener's views than it does about Hitler's. Hitler consistently subordinated his socialism to his evolutionary ethic, while for Wagener the socialist agenda was paramount. Wagener is also one of the few to report that Hitler specifically mentioned Darwin when discussing natural selection. According to Wagener, after explaining the survival of the stronger and better, Hitler stated, "Selection therefore runs a natural course. As Darwin correctly proved: the choice is not made by some agency—nature chooses."[25] Whether or not Wagener remembered correctly that Hitler specifically mentioned Darwin, he clearly recognized the Darwinian character of Hitler's ideas.

Other colleagues of Hitler's confirm Dietrich's, Junge's, and Wagener's observations. According to his adjutant, Nicolaus von Below, Hitler addressed army generals and field marshals on January 27, 1943. At the end of his talk he told them, "I have no other desire than to comply with this law of nature that stipulates that only the one who struggles for this life and is prepared, if necessary, to risk his own life for it, will gain his life."[26] Another close colleague of Hitler, Hans Frank, who served as Hitler's personal lawyer before 1933 and became governor of occupied Poland in 1939, admitted that Hitler often told him and others that war was an inescapable part of humanity. Hitler depicted nature as a constant struggle for sustenance and living space. Then he would comment that natural laws cannot be evaded, so anyone trying to forsake the struggle is pursuing an unrealistic dream. He contemptuously dismissed these idealistic dreams as "pacifistic twaddle." Frank's remonstrations that he considered Hitler's ruminations merely theoretical and did not think Hitler was making concrete plans for war shows either Frank's mendacity after the fact or naiveté earlier. In either case, Frank's own statements about Hitler's philosophy of struggle are remarkable for their forthrightness, since they show that Frank should have known where Hitler's philosophy would lead.[27]

The Struggle for Existence and Morality

Hitler often described the human struggle for existence as a pitiless form of competition, full of brutality and death. It took no cognizance of human moral standards. He regularly chided peace activists as naïve, since they hoped to escape from the laws of nature into an idyllic, but impossible, peace. They failed to appreciate that the struggle for space has been going on for innumerable epochs and will continue without ceasing into eternity—or at least as long as organisms continue to exist. Hitler admitted that this struggle was not pleasant, but he did not think it could be avoided. Atrocities were inevitable parts of these human conflicts, but they brought advance to those who ultimately triumphed. He stated that "humans have become masters over other beings through an inexorable struggle, yes a seemingly cruel struggle, a war of extermination with the goal of subjugating others."[28] This philosophy of cruel struggle would steel Hitler to commit unspeakable atrocities, all of which he explained as natural events caused by unavoidable natural forces.

He explained the pitiless character of this struggle extensively in a 1928 speech focusing on the human struggle for existence. He portrayed the struggle for existence as a universal process leading to the victory of the stronger and thus producing a "higher breeding." He then asserted that the struggle for existence consisted not only in humans competing with other organisms, but it also pitted people against other people. He reminded his audience that this struggle is not pleasant:

> This struggle occurs down to the lowest organisms; innumerable species have been defeated and exterminated, while others are poised on the brink of this destiny. Do you think that with humans this should be otherwise? Then, where does the boundary between the lowest New Zealand native, the Bushman, the tree climber and the ape lie? Where practically is the boundary here?[29]

The point here is obvious. Humans are subject to the same struggle for existence that leads to the extermination of other organisms. We cannot escape from nature. Immediately after making this point, Hitler told his audience that humans are not equal, a point implicit in the quotation above, where Hitler called into question the boundary between apes and those races he considered inferior.

In the same speech, after this discussion of the struggle for existence, Hitler spelled out its implications for morality: "On this earth the right of the stronger reigns, the right of struggle and the law of victory; but if you think that righteousness reigns, you are deceiving yourself."[30] Hitler expressed this idea—that might makes right—many times in many different ways throughout his life. In a 1923 speech he stated:

> Decisive is the power that peoples (*Völker*) possess; it shows that before God and the world the stronger has the right to accomplish his will. From history one sees that right in itself is useless, if behind it does not stand a mighty power. Whoever does not have the power to accomplish his right finds the right alone completely useless. The strong always triumph... All of nature is an unceasing struggle between strength and weakness, a constant victory of the strong over the weak.[31]

In a speech several years later, Hitler noted that the Nazi Party is "a fellowship of fighters and of hatred." Then, after explaining that this struggle necessarily includes imperialism, he turned his attention toward his internal enemies—the Jews. As he was wont, he called them parasites, a trope that dehumanized them and aroused repugnance. Then, he menacingly provided another analogy from nature. We cannot blame a tiger if it kills a person, he remarked, since this is merely a natural event. However, this does not mean we need to allow it to kill us, for the "right of the struggle" applies here.[32] By comparing his enemies with parasites and beasts of prey, Hitler could justify harsh measures as self-defense in the struggle for existence.

Clearly, Hitler thought this "right of the struggle" trumped all moral standards. He stated this concretely in a speech in January 1932 to the Düsseldorf Industry Club. He told the gathered businessmen,

> Politics is and can be nothing other than the realization of the vital interests of a people and the practical waging of its struggle for life with all means available. Thus it is quite clear that this struggle for life has its initial starting point in the people itself, and that at the same time the people is the object, the value in and of itself, which is to be preserved. All of the functions of this body politic should ultimately fulfill only one purpose: securing the preservation of this body in the future.[33]

Here Hitler indicated that the sole purpose of politics is to advance the cause of the *Volk* in its struggle. He was thereby justifying any policy that assisted the German *Volk* in their competition with others. He also explained in this speech that neither foreign policy nor economics had a higher priority. Both were means to an end, which was victory in the struggle for existence.

Hitler measured the righteousness of moral standards primarily by success in the struggle for existence. In *Mein Kampf* he summed up this point of view: "Every world-moving idea has not only the right, but also the duty, of securing those which make possible the execution of its ideas. Success is the one earthly judge concerning the right or wrong of such an effort."[34] However, Hitler then proceeded to qualify this stress on success slightly by claiming that short-term success is not the judge. Merely attaining power, as the Weimar Republic did in 1918–1919, is not success. Rather, he asserted, success must be measured by its impact on the *Volk*. Thus Hitler meant that long-term success that furthers the survival and reproduction of the *Volk* is the paramount value. However, if indeed Hitler's ethical philosophy can be summed up by the maxim, "might makes right," my contention that Hitler followed some kind of coherent ethic makes no sense. Indeed, at first glance Hitler's attempt to embrace nature's order, complete with its death, destruction, and brutality, seems rather amoral.

Indeed, though Hitler's vision of human struggle did sweep away most traditional forms of ethics, including Christian, Kantian, and Utilitarian, his ethical philosophy was not completely amoral. Rather, two factors in Hitler's worldview kept him from completely descending into nihilism. First, Hitler conceived of the struggle for existence itself as a good thing, because it promoted biological advance. Thus for Hitler evolutionary progress became the highest arbiter of morality. He expressed this clearly in many speeches and writings, including his *Second Book*, where he stated, "Therefore, ideals are healthy and appropriate as long as they help to reinforce a people's inner and collective strength, so that these forces can contribute in carrying out the struggle for life."[35] For Hitler, then, the way to discover if a moral ideal is correct is to ask: Does it benefit the individual or the *Volk* or the race in the struggle for existence? Does it advance the evolution of humanity or does it lead to biological degeneration?

Second, Hitler believed that evolution had produced morality, which marked an advanced stage of human evolution. This idea is admittedly somewhat contradictory to the previous idea that only success in the struggle for existence defines what is morally good. If morality is merely the product of human evolution, how could the process have any transcendent value? How can there be a "higher morality" at all? Hitler clearly did not believe that morality has any transcendent existence beyond humans' own experience, for in *Mein Kampf* he stated that, except for purely logical constructs, all human ideas—and here he explicitly included ethical ideas—are tied to human existence. If those humans who uphold a particular idea— whether all of humanity or just one race—perish, the ideas vanish with them, according to Hitler.[36] It is crucial to understand this point to interpret Hitler's ethical views, for he clearly opposed any transcendent, universal, or objective moral standards. He also seemingly rejected any overarching philosophical idealism, such as Platonism or Hegelianism, that gave primacy to ideas. Hitler was by no means alone in clinging to these contradictory principles. Many thinkers embracing evolutionary ethics in late nineteenth and early twentieth-century Germany likewise believed that morality was the product of naturalistic, evolutionary processes, but they also believed that the process itself defined what is moral.[37]

Hitler's Belief in Human Evolution

Hitler spoke and wrote incessantly about evolution, natural selection, and the struggle for existence, especially the struggle between races. It should be patently obvious from these discussions that he believed in human evolution. Though he discussed the evolution of races more than the evolution of humans from animals, at times he did explicitly discuss the animal origins of humans. While almost all scholars recognize that Hitler was a social Darwinist and thus embraced human evolution, a few people wrongly think that Hitler rejected evolution. This misconception ignores the vast preponderance of evidence and is based on a single passage in his Table Talks, where he expressed reservations about human evolution.[38] During this private conversation in January 1942 he reportedly stated:

> Where do we get the right to believe that humanity was not already from its earliest origins what it is today? Looking at nature teaches us that in the realm of plants and animals transformations and further developments occur. But never within a genus has evolution made such a wide leap, which humans must have made, if they had been transformed from an ape-like condition to what they are now.[39]

There are several problems with placing much weight on this one statement. First, if one examines the context, Hitler prefaced these remarks by stating that he was currently reading a book about the origins of human races. It thus seems likely that the opinions he expressed at this particular time were heavily colored by his current reading. They were certainly not long-standing views of his. Never before or later did Hitler make any statements denying or doubting human evolution. Second, while his offhand remarks do admittedly call into question human evolution from animals, he simultaneously clearly confessed belief in evolution for all other organisms. Third, and most important, many times earlier and at least twice afterward, Hitler clearly expressed his belief in human evolution. We have already examined many passages where he mentioned the "higher evolution" of humans through the struggle for existence. This is pretty strong proof in itself, but as I will show, Hitler was even more explicit at times about his belief in human evolution. He also often remarked that humans are ineluctably a part of nature and cannot escape from the same laws of nature governing everything else.

One line of evidence suggesting that Hitler's statement in January 1942 should not be given much weight was that it completely contradicted many earlier statements by him, where he minimized the distinction between the "inferior" human races and animals. The distinguished historian Gerhard Weinberg noted that a "significant facet of [Hitler's] racialist doctrine was its rejection of the biblical distinction between man and other creatures."[40] In his closing speech at the Nuremberg Party Rally in 1933, Hitler stated, "The gulf between the lowest creature which can still be styled man and our highest races is greater than that between the lowest type of man and the highest ape."[41] This point had been made repeatedly by Ernst Haeckel and other German Darwinists in their attempts to make human evolution plausible. In a 1927 speech, while discussing the importance of the struggle for existence for humanity, Hitler claimed that "the boundary between human and animal has been drawn by humans themselves." He then argued that the Aryan race was responsible for all major advances in technology and culture. "Humanity owes everything great to struggle and to a race, which has triumphed. Take away the Nordic German, and then all that remains is ape dances."[42] His statement that humans have created the animal-human boundary, together with his discussion of racial differences among humans, clearly implies that he was wanting to redraw this boundary. In his view only Nordic people have ascended culturally above the apes. Comparing "lower" races with apes to dehumanize them was a common trope widely used not only by Hitler but also in a good deal of Nazi propaganda.

Goebbels in his diaries reported a conversation with Hitler on December 29, 1939, which confirms that Hitler considered humans not all that far removed from animals. Just after mentioning Hitler's vegetarianism, he stated that Hitler "did not think much of Homo sapiens." Hitler told Goebbels that humans should not consider themselves so exalted. Though many think that we alone possess reason, speech, and a soul, how do we know that other animals do not also have these, Hitler asked. Though Goebbels did not specifically mention the animal origin of humans in this conversation, he did make clear Hitler's low view of humanity.[43] This is directly contradictory to the view Hitler expressed in January 1942 about humanity's position far above the animal kingdom. Once again, most often Hitler stressed the proximity of humans and animals.

An even stronger piece of evidence that Hitler believed in human evolution was a statement he made in a 1927 speech. After emphasizing the importance of the "law of the eternal struggle," he told pacifists,

> You are the product of this struggle. If your ancestors had not
> fought, today you would be an animal. They did not gain their
> rights through peaceful debates with wild animals, and later per-
> haps also with humans, through the comparative adjustment of
> relations by a pacifist court of arbitration, but rather the earth has
> been acquired on the basis of the right of the stronger.[44]

This is a clear indication that Hitler believed both in the animal ancestry of humans and in the role of the struggle for existence in advancing
human evolution.

During his Table Talks, Hitler also strongly professed belief in evolution. On October 24, 1941, he told his dinner guests that the church's
doctrine of creation from the Bible was in complete contradiction
with the theory of evolution. He claimed that as a school boy he had
already recognized the contradiction between what he was learning
in his religion class and his science class. He then proceeded to criticize Christianity, and lamented that contemporary discussions of the
science-religion nexus were far behind that of Enlightenment thinkers.
He specifically mentioned Voltaire and Frederick the Great as deep
thinkers about religion, showing his disdain for organized Christianity.
He then stated that science was making great strides and would ultimately supplant the church's doctrine: "Next to the gigantic power of
scientific research the dogma [of the church] will one day grow pale."[45]
In the science-religion conflict Hitler clearly was taking the side of
science and evolutionary theory against religion and the church. He
underscored this once again a few weeks later, when he stated, "Today
no one who is familiar with natural science can any longer take the
doctrine of the church seriously."[46] For Hitler science, especially evolutionary biology, clearly took priority over religion.

This is even clearer when he steered the discussion toward human
evolution. At the end of this lengthy monologue on evolution, science, and religion, he unequivocally expressed belief in the theory of
human evolution by stating, "There have been humans at the rank
at least of a baboon in any case for 300,000 years at least. The ape is
distinguished from the lowest human less than such a human is from a
thinker like, for example, Schopenhauer."[47] Hitler provided this ring-
ing endorsement of evolutionary theory, including human evolution,
just a couple of months before the conversation expressing doubt about
human evolution. Hitler's secretary Junge also remembered that Hitler

believed in human evolution. She reported that during one of his discussions about human evolution, Hitler remarked that scientists were not certain about the exact ancestors of the human species, but they had certainly evolved from reptiles through mammals, and possibly through apes.[48]

In addition to all these statements explicitly stating his belief in the animal ancestry of humans, Hitler often implied that humans evolved through the struggle for existence. In his *Second Book* he remarked that world history before humans appeared was a clash of geological forces. Long before the advent of humanity nature was filled with conflict. Humans arrived late in earth history, he explained, and though he did not explain in detail how this happened, he does describe it as evolution through struggle. After their appearance, humans have had to engage in "a never-ending battle of humans against animals and also against humans themselves."[49] In a separate passage in his *Second Book* he explained that just as the earth experienced geological transformations and just as some organisms go extinct, while others evolve, so the possession of land by peoples changes historically. Anyone who unrealistically tries to end this struggle for land among humans would "thereby also eliminate the highest driving force for their own evolution."[50] In *Mein Kampf* he ridiculed those who thought they could escape "the iron logic of Nature" with its universal struggle for existence. Humans owe their very existence solely to this natural struggle, he stated, and those who try to opt out of the struggle will only seal their own doom.[51] This confirms again his belief in the evolutionary origin of human beings.

At least twice after expressing skepticism about human evolution in January 1942, Hitler reaffirmed his belief in human evolution. Less than two months afterward he claimed that men shaving off their beards is "nothing but the continuation of an evolution that has been proceeding for millions of years: Gradually humans lost their hair." While this statement is ridiculous, it clearly expresses Hitler's belief that humans had evolved from hairy animals over millions of years. Finally, less than a year before he died, Hitler again professed belief in human evolution. In a secret speech to generals and military officers in June 1944, Hitler claimed that humanlike organisms had only existed for a few million years and humans for only about 300,000 years. This statement, together with the others discussed above, make very clear that Hitler believed that humans had evolved from apelike ancestors and that human evolution was still occurring. The context of this statement is also very revealing, for this secret speech is remarkable for its overt

Darwinian themes and its explanation of how these principles relate to ethics. In the speech's first sentence Hitler remarked that war is an inevitable phenomenon, and then he continued:

> Nature teaches us with every look into its working, into its events, that the principle of selection dominates it, that the stronger remains victor and the weaker succumbs. It teaches us, that what often appears to someone as cruelty, because he himself is affected or because through his education he has turned away from the laws of nature, is in reality necessary, in order to bring about a higher evolution of living organisms.[52]

If this was not enough to make clear that he thought humans were inextricably entangled in the web of evolution, natural selection, and the struggle for existence, Hitler then applied these principles forthrightly to humans. People cannot escape these natural laws, since "we humans have not created this world, but rather we are only very small bacteria or bacilli on this planet." Such was Hitler's view of the significance and dignity of humanity in light of evolutionary processes.

Hitler then warned these officers against practicing humanitarian ethics, since this would condemn humans to extinction, as other species would outcompete and supplant us. A short time later in the same speech he spelled out the implications of the evolutionary process for ethics. He stated, "War is thus the unalterable law of all life, the precondition for the natural selection of the strong and simultaneously the process of eliminating the weaker. What appears to people thereby as cruel, is from the standpoint of nature obviously wise." Nature does not care about any abstract human rights, but judges solely according to the right of the strong, he explained.[53] Human evolution was thus clearly central to Hitler's vision of ethics, politics, and history.

Some critics might object that Hitler in *Mein Kampf* sometimes claimed that humans were created in the image of God. In one memorable passage Hitler wrote that marriage should be "an institution which is called upon to produce images of the Lord and not monstrosities halfway between man and ape." A few pages later he asked if we should not "put an end to the constant and continuous original sin of racial poisoning, and to give the Almighty Creator beings such as He Himself created?"[54] Are not these statements clear evidence that Hitler rejected an evolutionary origin for humanity?

Not really. Many theistic evolutionists, both then and today, believe that God created humans in his image through the process of evolution. Even if Hitler were a sincere believer in God and not merely using God-language for propaganda purposes, this would not imply that he rejected evolution. In none of the relevant quotations from *Mein Kampf* does Hitler state that humans were specially created in the recent past by the miraculous intervention of God. On the contrary, Hitler repeatedly insisted that humans are subject to inescapable natural laws and that they are the product of eons of change. He often presented evolution as a universal process encompassing humans as well as other creatures.

Another reason that Hitler's allusions in *Mein Kampf* to humans as images of the Lord does not count as clear evidence against his evolutionary views is because in these passages Hitler's ideas seem to derive from, or at least parallel, that of the Aryan racial theorist Jörg Lanz von Liebenfels, who definitely believed in Darwinism. In the first of the passages I quoted above, Hitler referred to "monstrosities halfway between man and ape," and in the second he called mixing races the "original sin." Two decades before Hitler penned these words, Lanz von Liebenfels taught that the original sin was Eve copulating with an animal, thus producing a race that was half-man, half-ape. Belief that humans fell from a pristine original state may not seem particularly consistent with evolutionary thinking, but Lanz von Liebenfels (and also Hitler) accepted these seemingly contradictory strands of thought. Lanz von Liebenfels clearly embraced Darwinism and interpreted the Bible in an evolutionary sense. He claimed in one article that "Moses is thus actually a Darwinist, yes even a modernist, since evolution and selection are for him the driving forces of all being."[55] However, both Lanz von Liebenfels and Hitler believed that the evolutionary process was not always characterized by progress. Degeneration was also possible, and race mixing was the chief culprit. Thus, they wanted humans to intervene in the evolutionary process to counteract the forces of biological decline.

Another way we know that Hitler's remarks in *Mein Kampf* were not denying human evolution is because in other passages of *Mein Kampf* Hitler forthrightly discussed evolution. In one passage Hitler discussed the distinction between humans and animals explicitly: "The first step which outwardly and visibly removed man from the animal was that of invention." These inventions aided humans in their struggle for life. This strongly implies human evolution, and the following paragraphs confirm this point. Hitler continued by explaining that while many people see primitive inventions as mere instincts, in reality they must have originated through creative individuals, or what Hitler called the force of personality:

> For anyone who believes in a higher evolution of living creatures must admit that every expression of their life urge and life struggle must have had a beginning; that *one* subject must have started it, and that subsequently such a phenomenon repeated itself more and more frequently and spread more and more, until at last it virtually entered the subconscious of all members of a given species, thus manifesting itself as an instinct. This will be understood and believed more readily in the case of man.[56]

Thus, in the midst of his discussion of humanity, he confirmed his belief in evolution and provided an account of the origin of instincts, especially human instincts. In Hitler's view instincts were not implanted in species by a creator, but they evolved in order to benefit organisms, including humans, in the struggle for existence.

This is not the only passage in *Mein Kampf* clearly articulating belief in human evolution. In the opening pages of his chapter on "Nation and Race," Hitler discussed the implications of racial mixing, especially in humans, for evolution. He argued that stronger races must dominate and not blend with weaker ones; otherwise "any conceivable higher evolution of organic living beings would be unthinkable." A few paragraphs later he stated, "No more than Nature desires the mating of weaker with stronger individuals, even less does she desire the blending of a higher with a lower race, since, if she did, her whole work of higher breeding, over perhaps hundreds of thousands of years, might be ruined with one blow." It is apparent from this and other passages in *Mein Kampf* that—whatever his religious views—Hitler clearly believed in human evolution.

The Evolution of Morality

Not only humans but morality itself was a product of the evolutionary process, in Hitler's view. In *Mein Kampf* Hitler denied that morality was transcendent, universal, or absolute. He attacked those who upheld a humanitarian ethic, because they tried to apply their ethic to all humanity. Hitler, on the other hand, vigorously denied that any ideas—and he specifically mentioned ethical conceptions—exist apart from the humans bearing these ideas. He rejected absolute moral standards, insisting that they are dependent on human ideas. At the same time, because of his racial inegalitarianism, he did not think that all races had morality, or certainly not the same morality.[57] Further, as we have already seen, Hitler relativized morality by appealing to the struggle for existence, which became a higher standard than any abstract idea about morality or humanitarianism. Hitler summed these ideas up by stating,

> When the nations on this planet struggle for existence—when the question of destiny, "to be or not to be," cries out for a solution— then all considerations of humanitarianism or aesthetics crumble into nothingness; for all these concepts do not float about in the ether, they arise from men's imagination and are bound up with man. When he departs from this world, these concepts are again dissolved into nothingness, for Nature does not know them.[58]

Morality is thus not a set of transcendent, timeless, and universal principles, but rather a contingent characteristic of (some) humans.

How did morality originate in the first place, then? This is not a topic Hitler broached very often, but in a 1920 speech, "Why We are Anti-Semites," he provided some important clues. In this speech Hitler began by proclaiming that the key characteristic dividing humans from animals is labor. This point, which was quite similar to the view that the socialist leader Friedrich Engels promoted, was not particularly controversial. However, because Hitler later distinguished between races that labor (Aryans) and those who do not (Jews), his distinction would be inflammatory. Labor originated, according to Hitler, in the harsh northern climes. Humans were compelled "to struggle practically for their existence" against the elements. They had to expend considerable effort to gain their sustenance. By contrast, in the more hospitable areas of the earth, humans had an easier life. They could find food with far less effort. Because of this, the people of the north acquired the duty to labor, while those in more favorable climates did not.[59]

Though the duty to labor does not capture all aspects of morality, nonetheless it does seem to serve a fundamental role in Hitler's vision of morality. The upstanding German who diligently earned a living by his own labor was the epitome of responsibility and moral character. Further, Hitler claimed that because they were committed to labor as a social duty, Nordic men alone were able to found and develop political states. All major world empires and civilizations had been established by Aryan peoples, according to Hitler and like-minded racial theorists.[60]

Another reason that the northern climate contributed to the evolutionary advance of the Nordic race, according to Hitler, was because "the unprecedented necessity and frightful privation worked as a means for racially pure breeding." The weak and sickly quickly perished, leaving those who were healthy and vigorous to propagate the race. This ensured that the Nordic race would be physically superior to other races, because more favorable climates would allow weaker individuals to survive and reproduce.[61] Hitler's stress on the influence of the climate in shaping Nordic biological characteristics—including moral traits—was a staple among Nordic racists in early twentieth-century Germany.

In this speech, as well as in many others, Hitler stated or implied that morality was a tool to help win the struggle for existence, either against the harsh elements or against fellow humans. An innate sense of morality was a biological instinct that contributed to the survival of the species. Human moral standards, then, were only a means to an end. It was the goal—survival and propagation of the species—that was supremely important for Hitler. Morality was only important when it served those ends. If it hindered survival and reproduction, so much the worse for morality.

CHAPTER THREE

Racial Struggle

Hitler's Scientific Racism

When the leading Nordic racial theorist Hans F. K. Günther strode to the podium to deliver his inaugural address as professor of social anthropology at the University of Jena in 1930, his audience included none other than Adolf Hitler, a true comrade in racial ideology. The event was so important to the Nazis that Hermann Göring showed up, too, but only for the dinner celebration after the lecture.[1] Günther embraced a racial worldview blending elements drawn from Darwin, Gobineau, and other scientists and racial theorists. Imbuing his fellow Germans with Günther's Nordic racism was a high priority with Hitler, and indeed he and his party were instrumental in placing Günther in his professorship. Earlier in 1930 the Nazis had formed a coalition cabinet in the German state of Thuringia with the Nazi leader Wilhelm Frick as Minister of Education. In February 1930 Hitler wrote to an unknown correspondent that one of Frick's first responsibilities would be to establish a chair in racial studies (*Rassenkunde*) at the University of Jena. He hoped Günther would occupy the new position.[2] Frick carried out the Führer's will, appointing Günther over the objections of the faculty. In 1935 the Nazis would elevate Günther even further by appointing him professor of social anthropology at the prestigious University of Berlin.

Later, in 1930, Hitler thanked one of the leading anti-Semitic racial theorists, Theodor Fritsch, for sending him a copy of the thirtieth edition of his book, *Handuch der Judenfrage* (*Handbook on the Jewish Question*). Whether Hitler actually studied Fritsch's book when he lived in Vienna, as he claimed in this letter, is open to question. However, he certainly knew about Fritsch by the early 1920s in Munich, since the Nazi newspaper, the *Völkischer Beobachter*, occasionally ran notices of Fritsch's journal, *Hammer*, which was one of the leading organs for anti-Semitic racism in Germany at that time. In 1925 Fritsch had sent Hitler a copy of another anti-Semitic book he wrote.[3] Hitler's praise for Fritsch was effusive: "I am convinced," he stated, "that this [book] worked in a special way to prepare the ground for the National Socialist anti-Semitic movement. I hope that other editions will follow the thirtieth edition and that the book will gradually come to be found in every German family."[4] Already a year earlier, in an article celebrating the ten-year anniversary of his joining the Nazi Party, Hitler had acknowledged Fritsch as a pioneer in fighting the Jews.[5]

Racial theory was fundamental to Nazi ideology, and it became one of the leading features of their policy once they took control of Germany. From Hitler's earliest writings and speeches to his final testament, from the first Nazi laws discriminating against the Jews in 1933 to the Holocaust, race was a foundational principle of Nazi ideology and a factor motivating or influencing almost every Nazi policy. In the beginning of volume two of *Mein Kampf*, Hitler summed up his racial worldview, stating that

> the folkish philosophy [i.e., Hitler's own view] finds the importance of mankind in its basic racial elements. In the state it sees in principle only a means to an end and construes its end as the preservation of the racial existence of man. Thus, it by no means believes in an equality of the races, but along with their difference it recognizes their higher or lesser value and feels itself obligated, through this knowledge, to promote the victory of the better and stronger, and demand the subordination of the inferior and weaker in accordance with the eternal will that dominates this universe. Thus, in principle, it serves the basic aristocratic idea of Nature and believes in the validity of this law down to the last individual. It sees not only the different value of the races, but also the different value of individuals.[6]

Two important principles in this passage would be central for Hitler's ideology and policies: the inequality of races, and the eternal struggle for existence between races, which ultimately leads to further evolution of the higher race and the submission—elsewhere he would say death, destruction, or annihilation—of the inferior races. As he indicated

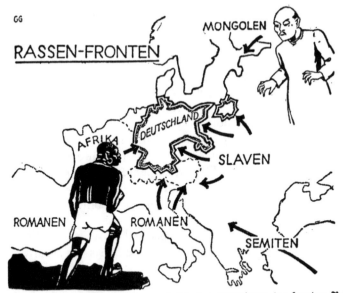

Nur mit der eigenen Volkskraft ist es möglich, das Vordringen der fremden Rasse
abzuwehren.

Figure 3.1   "Racial Fronts" (from a book by J. W. Ludowici, a Nazi official)

here, the whole purpose of the state (and of Nazism) was to advance
the interests of the best, strongest, and most valuable race—which, of
course, he identified as the Aryan race—in this struggle.

Racial inequality predated Darwinism by centuries, and Gobineau
wrote his major work on *The Inequality of the Human Races* before
Darwin published his theory. However, as many scholars have noted,
Darwin and many Darwinian biologists in the late nineteenth and
early twentieth centuries integrated racial inequality into evolutionary theory in ways that transformed and intensified racism.[7] They also
provided a scientific justification for racism, furthering the popularity
of scientific racism in the late nineteenth and early twentieth centuries. Darwinian theory required that significant variations must exist
within species; otherwise there would not be anything for natural
selection to select. Stressing racial inequality thus served an important function in Darwinian theory, because races could be construed
as incipient subspecies or even species. Indeed the leading Darwinist
in Germany, Ernst Haeckel, argued that human races were so different
that they constituted ten or twelve separate species. Some Darwinian

anthropologists even claimed that different human races had evolved from different simian species.

Further, evolutionary theory also required that there be a continuum between different species. Racism helped Darwin and his contemporaries bridge the gap between simians and humans, because they thought Australian aborigines, black Africans, and other allegedly inferior races were far inferior to Europeans mentally and morally. They had simply not evolved as much as Europeans and were thus living intermediaries between simians and Europeans.[8] Almost all Gobineau's disciples in early twentieth-century Germany interpreted Gobineau through Darwinian lenses. Leading Darwinists, such as Haeckel, praised Gobineau, and Eugen Fischer even asked the leader of the Gobineau Society for a portrait of Gobineau to display at the entrance of his anthropological institute.[9]

The historian Benoit Massin explains how this Darwinian version of racism affected many German anthropologists in the early twentieth century:

> And for those embracing the new Darwinian approach in German anthropology, the implications of racial evolutionary hierarchies were even more radical: the replacement of the previous humanitarian ethics by a biological and selectionist materialism more concerned with the inequalities of evolution than the universal brotherhood or spiritual unity of humankind.[10]

Hitler clearly embraced this Darwinian version of racial inequality that viewed races as having evolved in varying amounts from their simian ancestors. He also drew from this inegalitarian viewpoint the same antihumanitarian ethical conclusions as the German anthropologists Massin analyzes.

Further, Darwinism would transform racial thought in the nineteenth century by introducing the idea that the racial struggle for existence helped produce biological progress. Gobineau was a pessimist warning against racial degeneration. In *The Descent of Man*, however, Darwin portrayed racial struggle as an important element bringing advancement to the human species by eliminating allegedly inferior races. He stated, "At some future period, not very distant as measured by centuries, the civilised races of man will almost certainly exterminate and replace throughout the world the savage races."[11] Though he did not advocate violence and killing to help evolution along, he still thought the elimination of the inferior "savage" races was a beneficial process.

So did many German racial theorists in the early twentieth century, including Haeckel, Woltmann, Theodor Fritsch, Eugen Fischer, Alfred Ploetz, and Fritz Lenz.

Hitler considered it one of his chief tasks to implement racial policy based on scientific racism.[12] He continually appealed to the laws of nature to justify his inegalitarian racist program. In his *Second Book* he remarked, "It will be the duty of the National Socialist movement to transfer the either already existing or future findings of scientific insights of racial theory—as well as the world history it elucidates—into practical, applied policy."[13] For Hitler racial policy—important as it was—was still only a means to an end. Ultimately, Hitler saw it as a way to triumph in the racial struggle for existence, which would lead humanity to ever higher stages of evolutionary development.

Racism was thus always in the service of evolution, which was the paramount value. Hitler continually stressed, both publicly and privately, that racial competition fostered evolutionary progress. In his chapter on "Nation and Race" in *Mein Kampf*, where Hitler set forth his racial views at length, he opened the chapter discussing the role race played in human evolution. This chapter was so central to Nazi ideology that it was the only part of *Mein Kampf* to be published as a separate booklet during the Nazi period. Five hundred thousand copies were printed, and Education Minister Bernhard Rust included it on a list of the 120 most important books for schools to acquire. Thus Hitler's views on race were widely distributed.[14] Hitler believed his ideas were rooted in the laws of nature, which humanity spurned at its peril. He scoffed at those who thought they could transcend nature and set aside its immutable laws, especially its racial laws. Trying to stymie nature in its course would only result in disaster, he explained, since "the man who misjudges and disregards the racial laws [of nature] actually forfeits the happiness that seems destined to be his. He thwarts the triumphal march of the best race and hence also the precondition for all human progress."[15] Conversely, racial awareness and policy based on it would foster improvement of the human species. In the first few pages of this chapter Hitler stressed the importance of racial inequality and racial struggle for his worldview.

Though anti-Semitism played an extremely prominent role in Hitler's racial views—as it did in this particular chapter of *Mein Kampf*—his racial arrogance was not directed solely at Jews. He considered the so-called Aryan or Nordic race—terms he used interchangeably—higher than any other race on the earth. He believed Nordic blood had been preserved most purely in the German and Scandinavian peoples. For

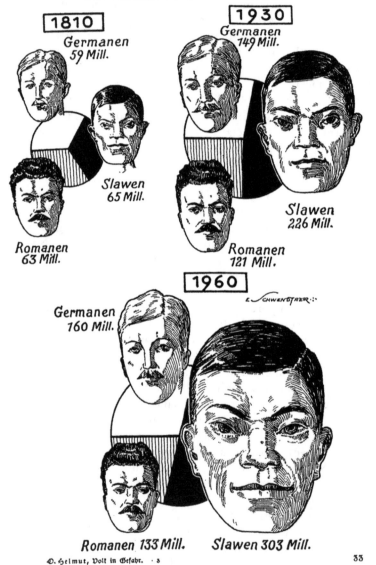

Figure 3.2 "Fertility and Race: The Growth of the Slavs in Europe" (from Volk in Gefahr)

Hitler this racial superiority of the German people implied that they were more valuable than other people. In a 1938 speech he claimed that historical development depends on the differing value of races, and then emphasized his own commitment to German supremacy: "The value of the German *Volk* is incomparable. I will *never* allow myself to be per-suaded, that any other *Volk* could have *more* value! I am convinced that our *Volk*, especially today, in its gradual racial improvement represents the highest value, that has ever been on the earth up to this time."[16] In 1935 he told a Berlin audience that the German *Volk* is everything. He exalted it to the supreme value.[17]

Whenever Hitler asserted the primacy of the German *Volk*, he was thinking of it as a racial category. He often used the terms German *Volk* and Aryan race interchangeably in *Mein Kampf* and elsewhere. He also regularly insisted that the *Volk* is defined by its blood, meaning its bio-logical, hereditary qualities.[18] In early 1922 Hitler proclaimed in a Nazi Party circular that the most fundamental principle is "namely that the essence and character of a *Volk* is not to be found in religion, nor is it to be found in language, but rather is primarily in the blood; the blood makes the race."[19] Later, in *Mein Kampf* he scoffed at the idea that one could Germanize black Africans or Chinese by teaching them German, since "nationality or rather race does not happen to lie in language but in the blood."[20] Hitler always used the term "blood" to refer to heredi-tary factors, so for him the *Volk* was determined by its biological traits passed on from generation to generation.

In Hitler's *Second Book* he continued to use the terms "race" and "*Volk*" interchangeably. This is apparent throughout the book, but is especially obvious in Chapter Five, which opens with Hitler proclaim-ing, "I am a German nationalist." He then explained that peoples of other races were not part of the German *Volk*. He specifically criticized those earlier nationalists who tried to integrate Czechs or Poles into the German national community. He explained that the National Socialist Party's "national conception will not be determined by previous patri-otic notions of state, but rather by ethnic and racial perceptions."[21] Thus for Hitler the German *Volk* was not culturally, but racially, defined. In another passage he asserted, "If we start from the premise that one *Volk* is not equal to another, then their value is not equal either." He then remarked that the greater the "racial value" (*Rassenwert*) of a *Volk*, the greater their "value of life" (*Lebenswert*).[22] The value of any *Volk*, then, was based on its racial composition.

Another clear illustration of Hitler's interchangable usage of *Volk* and race is a November 1929 speech on the Nazi worldview. Here Hitler elevated the German *Volk* to the highest value in the world. Whatever served the interests of the German *Volk* was morally right and whatever was detrimental to the German *Volk* was evil. Hitler first explained that the primary element of the Nazi worldview was that the strong should triumph and impose their views on the world. He then trotted out the example of the white race triumphing over other races throughout the world to demonstrate this principle. After explaining the importance of this racial struggle, he asserted, "A worldview is correct, then, when it can lead a people (*Volk*) upward." Here—as in all his speeches and writings—*Volk* and race were identical.[23]

Hitler explained his biological task to preserve and improve the German *Volk* in a 1925 leaflet, "The Social Mission of National Socialism." The goal of Nazism, he stated, is the "preservation and advancement, nourishment and security of our *Volk* and of the most valuable racial elements that are the basis of this *Volk*....For us the Aryan is the chief bearer of human culture, and our *Volk* has the advantage that it is still able today to characterize a large part of its blood as Aryan." His job was to help find ways to increase the Aryan blood and thereby improve the national body (*Volkskörper*).[24] This shows once again that for Hitler the German *Volk* was a biological entity, not a cultural construct.

The biological and racial content of the term *Volk* is important to grasp, because Hitler used the term *Volk* much more frequently than race, leading many people—especially his contemporaries—to think that nationalism was preeminent in Nazi ideology. Certainly it is true that Hitler's use of the term *Volk* resonated with German nationalists, but Hitler' form of nationalism was not the same as that of many other Germans. Most nationalists in the nineteenth century—and many in the early twentieth century—had defined the nation according to linguistic and cultural criteria. Hitler's racist vision of nationalism was by no means unique, as many leaders of the Pan-German League also embraced a racial definition of the *Volk*. However, by using the term "*Volk*" so frequently, Hitler was able to appeal to all nationalists, whether or not they defined the nation by racial criteria. Like all nationalists, Hitler was interested in forging a common language and culture for the German people. However, he only wanted to unify culturally those belonging to the Aryan race. Race was primary, and the nation must conform to racial boundaries.

In order to purge and then preserve the nation from racial aliens, Hitler proposed discriminatory measures against those who were not of Aryan ancestry. The Jews were the primary target of this racial discrimination, since they were considered the most acute threat to the Aryan race for several reasons: (1) They were the largest identifiable non-Aryan population in Germany; (2) many were prominent in business, medicine, and journalism; (3) they were allegedly immoral and criminal (see chapter 4); and (4) Jews were allegedly privy to a worldwide conspiracy aimed at defeating the Aryans. Hitler was a true believer in the international conspiracy theory popularized through the *Protocols of the Elders of Zion* in the early twentieth century.

In order to win the racial struggle against the Jews, the Nazi regime introduced anti-Semitic legislation within weeks of gaining power. It seems apparent that the Nazi regime generally moved cautiously in the mid-1930s in introducing discriminatory legislation and policies, because they feared international sanctions and boycotts. On April 7, 1933, the Civil Service Act removed Jews from government positions, including posts of professors and teachers. Later that year Goebbels set up the Reich Chambers of Culture, which effectively barred Jews from participating in the fine arts and journalism.

Hitler was especially interested in bringing citizenship laws into line with his racial ideology. In August 1920 Hitler told a meeting of National Socialists in Austria that some people in the German *Volk* were going hungry and some were emigrating because of economic privation. "As long as this is the case," he asserted, "we have a holy and moral right to demand that this Reich exist for our own ethnic comrades (*Volksgenosse*) and not for others."[25] His desire to define citizenship in racial terms was clearly articulated in an April 1922 speech, where he stated, "A citizen in the Reich that we want to build is he who is an ethnic comrade (*Volksgenosse*). And an ethnic comrade (*Volksgenosse*) is he who is of our blood."[26] These points had already been laid out in the Twenty-Five Point Nazi Party Program of February 1920. Point four demanded: "Only he who is an ethnic comrade (*Volksgenosse*) can be a citizen. Only he who is of German blood, regardless of his religion, can be an ethic comrade (*Volksgenosse*). No Jew, therefore, can be an ethnic comrade (*Volksgenosse*)." Thus, not only should blood or heredity determine one's *Volk*, but it should also determine eligibility for citizenship. This meant that Jews and other non-Aryans should be stripped of their citizenship. The Nazi regime never went quite this far until World War II broke out. In September 1935, however, Hitler announced new anti-Semitic laws at the party congress at Nuremberg. One of these, the Citizenship Law, effectively defined Jews as second-class citizens by setting up a new category of citizenship open only to Aryans. This did not quite fulfill the earlier proposal in the Nazi Twenty-Five Point

Program to strip the Jews entirely of their citizenship (Jews still had German passports, for instance), but it was a symbolic blow to the status of the Jews nonetheless.

In the official commentary of the Nuremberg Laws, Wilhelm Stuckart and Hans Globke, officials in the Ministry of the Interior, gave a scientific gloss to the discriminatory legislation. They stated,

> National Socialism opposes to the theories of the equality of all men and of the fundamentally unlimited freedom of the individual vis-à-vis the State, the harsh but necessary recognition of the inequality of men and of the differences between them based on the laws of nature. Inevitably, difference in the rights and duties of the individual derive from the differences in character between races, nations and people.[27]

The Nuremberg Citizenship Law and all racial legislation, then, were supposed to reflect racial inequalities rooted in nature.

Point five of the Nazi Party Program stipulated that all noncitizens, meaning all Jews and all non-Germans, should be treated as foreigners. Worse yet, point seven threatened deportation to all foreigners if the state is unable to support all its citizens. The following point demanded the deportation of all non-Germans who had entered Germany after the beginning of World War I. This was aimed largely at Eastern European Jews, one of the largest groups of recent immigrants. In 1938 the Nazi regime began implementing this by deporting all foreign Jews. By 1938 the Nazi regime also began forcing Jews to emigrate from Germany (even though no Germans were going hungry because of their presence). Those Jews arrested during the Crystal Night Pogrom in November 1938 were told to leave the country or else. By 1938–1939, Jews were forbidden to practice most professions and their businesses were forcibly "Aryanized." By the beginning of World War II, about 60 percent of Germany's Jews had fled the country.

Racial Superiority

Hitler hoped to redraw the boundaries of the German nation to include all those deemed racial comrades. This included most prominently those ethnic Germans living outside the German Reich. Thus, his goals comported quite nicely with Pan-German nationalists, who wanted all Germans to unite in a greater German Empire. Since there were significant populations of ethnic Germans in Austria, Czechoslovakia, and Poland, these were the first areas Hitler targeted for expansion.[28] However, he also wanted to bring ethnic Germans dispersed further afield back into the greater German Reich. After conquering Poland in 1939, the Nazi regime invited ethnic Germans from various Eastern European countries to come settle the newly annexed territory that had formerly been western Poland.[29] He also wanted to incorporate the Volga Germans into his greater Reich once he crushed the Soviet Union.

Interestingly, Hitler also desired to incorporate into the German *Volk* many people who did not speak German and did not consider themselves ethnically German. Norwegians, for instance, were considered fellow Aryans and were welcomed into the racial community, even though many of them hated the occupying Germans. Hitler issued a decree on July 28, 1942, stipulating that children fathered by German troops with Norwegian or Dutch mothers would receive special care to preserve the "racially valuable Germanic hereditary material."[30] This desire to incorporate Europeans from other nationalities into the Aryan racial community was reflected in the language used in racial legislation. One of the leading racial experts in the Interior Ministry, Arthur Gütt, admonished his fellow bureaucrats to avoid the term "German race," since no such race existed.[31]

Gütt apparently agreed with the Nazi racial theorist Günther, who had claimed that the German people were a mixture of several northern and central European racial types. Günther's views gained official approval during the Nazi period. However, his racial theories did not seem completely compatible with the notion of pure Aryan ancestry that so preoccupied Hitler, especially early in his career. It is unclear when Hitler embraced Günther's racial classification scheme, since he never discussed the racial composition of Aryans in sufficient detail. In a secret speech in May 1944, however, he did claim that Germans were a mixture of European racial elements, stemming most prominently from Nordic ancestry, but also including eastern (here he used Günther's term, "*ostische*") and Mediterranean racial elements.[32] It is likely Hitler had adopted Günther's racial views of Germans as a mixed race before 1930, when he pushed Frick to name Günther to a professorship in racial studies.

Hitler might even have altered his views on race in the 1920s under Günther's influence, though we cannot know for sure. We do know, however, that Hitler admired Günther, whose racial views underpinned many Nazi racial policies. In June 1933 the Nazi Minister of the

Interior, Wilhelm Frick, appointed Günther to the Expert Committee for Population and Racial Policy, which was responsible for drafting Nazi racial legislation. Further, official Nazi publications, such as the SS booklet, *Rassenpolitik*, relied heavily on Günther's racial ideas.[33]

Following Gütt's lead, by the mid-1930s many Nazi racial experts avoided the term Aryan altogether, and Hitler only rarely used the

Figure 3.3 Nordic racism in schools (from Nazi periodical)

term after coming to power in 1933 (though he did still use it occasionally). Thus, instead of "German race" or "Aryan," Gütt introduced the terms "German or related blood" to describe the "Aryans" in Nazi legislation and policy. This term was elastic and included most prominently Scandinavians, Anglo-Saxons, and the Dutch. Gütt explicitly excluded Jews, Gypsies, and blacks from the Germanic racial community, however.[34]

Nazi policies in the occupied territories of Eastern Europe also reflected Hitler's desire to promote the development of a supreme Aryan race that included more than just those ethnically German. In 1941 Hitler expressed the desire that Norwegians, Swedes, Danes, and the Dutch would settle in the Eastern occupied territories.[35] This is not so surprising, since Hitler considered the Scandinavians and Dutch fellow Aryans. However, more remarkable were the Nazi attempts to incorporate some Slavs into the German racial community. After conquering Czechoslovakia, the German Interior Ministry issued a regulation on March 29, 1939, to govern racial policy in Bohemia and Moravia. The Interior Ministry rules allowed some Czechs deemed racially superior to be assimilated into the German racial community, though it rejected all Jews, Gypsies, and members of non–European races.[36]

By 1940 Hitler had a fairly high regard for the Czechs, who had a vibrant industrialized economy, and he thought that perhaps half of the Czechs were racially valuable enough to assimilate into the German *Volk*.[37] On September 23, 1940, Hitler told Neurath that many Czechs could be assimilated into the German *Volk*, but "those Czechs who are racially useless and hostile to the Reich will be eliminated."[38] In an October 1940 meeting with Nazi officials in charge of the Protectorate of Bohemia and Moravia, Hitler approved a plan to allow some Czechs to join the German *Volk*, as long as they were properly screened to determine their racial fitness. He estimated that about half of the Czech population might qualify. He made clear, however, that neither "mongoloid" types nor the Czech intelligentsia could be Germanized. After taking charge in the Protectorate in September 1941, Reinhard Heydrich appointed racial experts to examine all applicants for German citizenship, including ones who had already been approved by his predecessors. As Chad Bryant has pointed out, however, Nazi administrators and racial experts had no coherent definition for what constituted racial fitness, so some officials were far more liberal than others in accepting Czechs into the German *Volk*.[39]

The Nazi regime followed a similar line in occupied Poland. In May 1940 Himmler called for the "re-Germanization" of any Poles,

Ukrainians, or other non-Jews in occupied territory who were deemed racially suitable. He declared that "the basis of our consideration must be to fish out of this mush the racially valuable, in order to bring them to Germany for assimilation."[40] On September 12, 1940, with Hitler's approval, Himmler issued a decree to compile German Ethnic Classification Lists in the occupied Polish territory. Polish citizens could be included on these lists if they had identified themselves as ethnically German before the beginning of the war, or if German authorities deemed them "capable of re-Germanization." Poles not included on the lists were given status as "protected subjects of the German Reich," while Jews and Gypsies were given no legal status whatsoever.[41] Just a few weeks later Hitler named Himmler the Reich Commissar for the Strengthening of the German *Volk*, which gave him considerable control over racial policies in the occupied territories.[42] The SS commissioned German anthropologists and other scholars to traverse Eastern Europe, identifying those people in the occupied territories whom they deemed racially valuable, regardless of their ethnicity.[43] Those designated as sufficiently Aryan were allowed to reside in areas set aside for German "colonization," while their fellow countrymen were deported further east to make way for newly arriving German settlers.

This propensity to include and incorporate non-Germans in the Aryan racial community illustrates yet again that ethnic and national identity were not crucial for Hitler, except perhaps as a means to unify Aryans in the racial struggle for existence. He did not care what language people spoke or whether they had read Goethe. The only pertinent consideration was whether or not they had what he considered superior biological traits. However, ambiguity and inconsistency dogged this enterprise, since neither Hitler nor the anthropologists and other racial experts assisting the Nazis in making racial determinations had a coherent, consistent way to discern racial fitness. All the racial terms of the Nazis—Aryan, German or related race, and so on—were ill-defined and highly subjective in application. Nonetheless, in theory Hitler's racism always trumped his nationalism. Nazi policies made a valiant effort to follow this ideology, but it was always bedeviled by the practical difficulty in categorizing people racially.

Most German proponents of Nordic or Aryan racism—including Hitler—believed that aside from the Scandinavians, Germans, Dutch, Anglo-Saxons, their descendants, and scattered other people having Aryan ancestry, all other races were inferior. In a 1932 speech, Hitler stressed the racial value of Germans and the importance of recognizing racial inequality. He scoffed at those who thought that "there is no essential difference in value between Negroes, Aryans, Mongols, and Redskins."[44] Indeed, Hitler clearly did not think any people of non-Aryan races had any value at all. In *Mein Kampf* he stated, "All who are not of good race in this world are chaff."[45]

Hitler had nothing but disdain for black Africans, whom he considered far below the level of Aryans. He admitted that blacks are humans, but still he considered them essentially different from and inferior to Europeans. In response to those who claimed that education could elevate blacks, Hitler protested in *Mein Kampf* that it is "criminal lunacy to keep on drilling a born half-ape until people think they have made a lawyer out of him, while millions of members of the highest culture-race must remain in entirely unworthy positions." Educating blacks, he remonstrated, is like training poodles. Hitler obviously considered blacks animalistic and less highly evolved than his beloved Aryan race. He also held out no hope for biologically improving blacks, since they were so far removed from Europeans. He scoffed that one could no more breed a black into a Scandinavian than one could breed a grasshopper into a rabbit.[46] He also criticized Christian missionaries from Germany, whose attempts at culturally improving black Africans

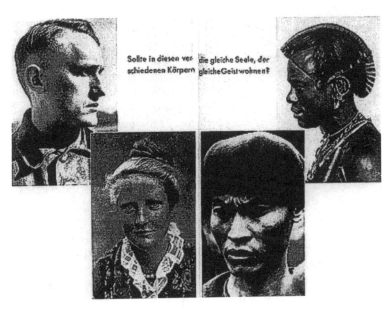

Figure 3.4 "Should the same soul, the same mind, inhabit these different bodies?" (from SS pamphlet)

would backfire by turning these "healthy, though primitive and inferior, human beings into a rotten brood of bastards."[47] Later he was also outraged that the United States sent blacks to compete in the 1936 Berlin Olympics. "The Americans should have been ashamed to let the blacks win medals for them," he told the Hitler Youth leader Baldur von Schirach.[48]

Black Africans were not the only race Hitler considered inferior. He remarked in *Mein Kampf* that it would be impossible to form alliances with natives in India or Egypt against their colonial lord Britain. "As a folkish man, who appraises the value of men on a racial basis," he stated, "I am prevented by mere knowledge of the racial inferiority of these so-called 'oppressed nations' from linking the destiny of my own people with theirs."[49] Just like many other racial ideologues in his time, Hitler believed that the European conquest and colonization of other parts of the globe was proof of the racial inferiority of the indigenous peoples, who simply did not have the inherent abilities—especially mental talent—of the European colonizers. European technological superiority supposedly reflected innate intellectual capacities lacking in non-Europeans. Hitler often remarked that the British were so successful in colonizing India because they did not racially mix with the inferior native population.

Despite his claim in 1923 that he would not form an alliance with the racially inferior natives of India or Egypt, he did, of course, form a military pact with Japan in 1936 and joined them in their war against the United States. Perhaps Hitler thought the Japanese were racially superior to Indians or Egyptians. After all, they had never been colonized, and they had appropriated European technology much more rapidly than most other non-Europeans. However, Hitler certainly did not think that the Japanese were on the same racial level as Germans. In *Mein Kampf* he specifically discussed the racial inferiority of the Japanese. According to Hitler, the only reason the Japanese had reached their advanced cultural level was because they emulated the Aryans and appropriated their achievements. Asians were imitators, not creative, he thought. On their own, they did not have the intellectual ability to develop an advanced culture.[50]

Hitler's alliance with Japan seems somewhat inconsistent with his racial ideology, and perhaps it is. Hitler was uneasy about it himself, confiding to Speer that he regretted siding with the yellow race against whites, but it was temporarily necessary.[51] While this attitude might seem like cynical Realpolitik, I still suspect that indirectly racial reasoning lurked behind his plans and purposes in his alliance with

Japan. Despite his misgivings about the inferiority of the Japanese, he acknowledged their ability to use European-developed technologies. This explained their military prowess, including their stinging defeat of Russia in 1905. Also, Hitler seems to have justified the alliance in order to help him defeat his racial enemies who were closer at hand, since he hoped—vainly as it turned out—that Japan would aid him in defeating the Soviet Union. This explanation seems especially likely in light of Hitler's reminiscences of his youthful response to the Russo-Japanese War. During that war he favored the Japanese, because he wanted to see Russia, the leader of the Slavic world, defeated.[52] By the post-World War I era, Hitler's hatred for Russia increased, for not only did he despise the majority Slavic population, but he viewed communism as one manifestation of a Jewish world conspiracy.

In 1939 Hitler justified his alliance with Japan by noting that a collapse of Japan "would not benefit European or other cultured nations, but would only lead to a certain Bolshevization of East Asia."[53] This statement may seem at first glance like the same old power politics of European imperialism, but for Hitler Europeans and cultured nations were a distinct racial group waging a racial war against the Jewish Bolsheviks. Thus, Hitler's statement implies that even in his dealings with Japan, he was using racial criteria to determine—or at least to justify—his foreign policy. Since the Japanese were no immediate threat to Germany in the racial struggle for existence, Hitler was willing to accept their help to fight his immediate racial enemies. This demonstrates once again that Hitler's tactics were flexible, but his ultimate goals were fixed.

Not only Africans and Asians, but many Europeans—especially Slavs—were also inferior to Germans, in Hitler's estimation. In *Mein Kampf* Hitler criticized earlier German diplomats for maintaining an alliance with Austria-Hungary before World War I. The problem with this alliance was not merely that Austria-Hungary had become progressively weaker and thus less valuable as an ally, but this weakness was linked to its racial composition. Hitler stated, "Also from the standpoint of racial policy, the alliance with Austria was simply ruinous. It meant tolerating the growth of a new Slavic power on the borders of the Reich."[54]

Elsewhere in *Mein Kampf* Hitler made clear his belief in Slavic racial inferiority. He claimed that the only reason Russia had become a great power was because the Russian elites had Germanic ancestry. They formed an aristocratic class dominating the inferior Slavs, who comprised the peasantry. The Bolshevik Revolution had eliminated most of the Aryan blood from Russia, thereby weakening it.[55] The Slavic elements in Russia, Hitler asseverated in his *Second Book*, do not even have the ability to organize a state.[56] Hitler's anti-Slavic racism was no secret. Stalin took note of it, informing the Seventeenth Party Congress in January 1934 that the Nazis believed "war should be organised by a

Figure 3.5   Slavs as Subhumans (cover of Nazi pamphlet)

'superior race,' say, the German 'race,' against an 'inferior race,' primarily against the Slavs."[57] Though Stalin ignored Hitler's primary obsession with the Jews, he rightly perceived Hitler's antipathy for Slavs.

Hitler's anti-Slavic racism led him to express contempt for the Poles in *Mein Kampf.* He criticized the policy of earlier German regimes that tried to Germanize the Polish populations living in Germany by requiring them to learn the German language. Hitler opined that if this kind of Germanization had succeeded, "the result would have been catastrophic; a people of alien race expressing its alien ideas in the German language, compromising the lofty dignity of our own nationality by their own inferiority."[58] Hitler clearly thought the Poles were a biologically inferior race that could not be elevated by education. In his *Second Book* he reiterated his view that Poles (and this time he included Czechs as well) could never be Germanized because of their inferior racial composition.[59] In later speeches Hitler accused Poles of ruining lands that had formerly been German. Poles simply did not have the racial qualities to maintain the high level of civilization that the Germans enjoyed, he supposed.[60] Depicting the Poles as an inferior race was not just war propaganda, though it surely served this purpose, too. In private conversations and in policy decisions, Hitler consistently spoke about and treated Poles as racial inferiors.

Once Germany occupied Poland and then later some parts of the Soviet Union, its occupation policies reflected anti-Slavic racism. John Connelly has argued cogently that Nazi racial policies toward Poles and other Eastern European Slavs was ambiguous and flexible. One reason for this was because they viewed some Slavs as racially valuable and assimilable into the German *Volk*, as we have already seen. Another reason was because the Nazis needed labor in areas where the Nazis could not bring enough German settlers to run the economy.[61] Nonetheless, Poles and other Slavs in occupied territories were exploited as slave labor precisely because they were deemed racially inferior. Thousands of Poles were sent to Germany as slave labor, and even the Poles remaining in Poland were essentially enslaved. The first instructions Hitler gave to Hans Frank as the governor of the occupied Polish territory known as the Generalgouvernement was to plunder Poland without mercy.[62] Himmler exemplified this attitude toward subject peoples in his October 4, 1943, speech: "Whether the other peoples live in comfort or perish of hunger interests me only in so far as we need them as slaves for our culture."[63] In 1942 Hitler told his associates that the Slavic peoples in the East should not receive education, hygiene, or medical treatment.[64] The Nazis introduced many discriminatory laws against

Poles, including segregating pubs and restaurants in the Wartheland, a western part of Poland that Germany annexed.[65]

Hitler's contempt for Slavs as racially inferior subhumans gave him supreme optimism that Germany would ultimately triumph over the Soviet Union. He consistently underestimated the military potential of the Soviet Union, because he always viewed them as racially inferior to the invading German troops. In December 1939 Hitler told Goebbels that he was happy that "Bolshevism has eliminated the Western European leadership class in Russia," since this weakened their nation.[66] A year later he told his generals he was confident that Germany could beat Russia in a military showdown, since "the Russian individual is inferior."[67] During the war he continued to exude confidence in victory because of the racial superiority of the Germans. In July 1942 Hitler told his colleagues that in the contest between different races the superior one will always triumph. It would be a violation of the laws of nature, he thought, if "the inferior would become lord over the stronger." He predicted that the Germans would become "the absolute lords over Europe."[68] Thus, his faith that Germany would eventually win the war, even when facing insurmountable obstacles, flowed in part from his view of German racial superiority.

Race and Culture

In an extended discussion of Aryan racial supremacy in *Mein Kampf*, Hitler claimed, "All human culture, all the results of art, science, and technology that we see before us today, are almost exclusively the creative product of the Aryan." If the Aryan perishes, human culture will gradually, but inevitably, die out.[69] The notion that Aryans were the sole creators of advanced culture because of their innate biological characteristics was widespread among racial theorists in the early twentieth century. None pushed this idea more vigorously than Woltmann, who incorporated the idea into many of his writings. He also wrote two books on the subject, one allegedly demonstrating that the Italian Renaissance was produced by Germanic people, and another arguing that the best of French culture had been produced by Germans. Hitler's ideas are also remarkably close to those that Erwin Baur, a leading geneticist, purveyed in *Germany's Renewal* in 1922. In an article, "The Downfall of the Cultured Peoples in the Light of Biology," he maintained that even if inferior races, such as blacks, were raised by European parents, they could not maintain the cultural level of Europeans, because they do not have the requisite racial qualities.[70]

In *Mein Kampf* Hitler divided the peoples of the world into three main categories: culture-producing races (or founders of culture), culture-preserving races (or bearers of culture), and culture-destroying races. Only the Aryans were culture-producing, and they alone had produced all human cultural achievements. The only concrete example of a culture-preserving race that he provided was the Japanese, who—he claimed—were able to utilize European technology, but unable to create it themselves. He obviously knew little or nothing about the rich history of Japan (or China), or about their splendid cultural achievements.[71]

Hitler discussed one culture-destroying race at length—the Jews. While Hitler's anti-Semitism by no means was derived from his evolutionary views, he—like Theodor Fritsch and many leading anti-Semitic publicists before him—did integrate his anti-Semitism into an evolutionary worldview. The biological form of anti-Semitism that came to dominate anti-Semitic racist views in the early twentieth century made it possible to integrate anti-Semitism and social Darwinism.[72] Hitler always insisted that Jews are not primarily a religion, but a race. Their biological, hereditary qualities defined them, and they were ultimately inferior to the Aryans. They possess a "tough will to live" and a strong instinct of self-preservation, and they have maintained a higher degree of racial purity than most other races, which accounts for their measure of success in history. However, Hitler insisted that they could not create culture on their own. At best, they could only imitate the culture of other higher races. Worse, they are parasitic, dragging down the cultural level of the people among whom they live.[73]

Hitler interpreted all of history through the lens of his Aryan racism. In a 1925 speech he explained his view of history succinctly:

We have no historical research that values the importance of the human races for the destiny of peoples (*Völker*). We need a conception of history, which views history not as just compiling a series of battles, but that penetrates into the racial instincts of conquest, the primal racial elements. Then comes the new worldview! History instruction must lead to the original factors: race and racial instincts.[74]

In Hitler's worldview, then, race was the secret to understanding history, just as the economy was central for Marxists. In *Mein Kampf* Hitler professed, "The racial question gives the key not only to world history, but to all human culture."[75] In a 1922 speech Hitler explicitly contrasted his philosophy with Marxism, "For us there is no class struggle, but rather racial struggle."[76] In Hitler's ideology the racial struggle was the force driving history forward, just as the class struggle was for Marxists.

In *Mein Kampf* he sketched out the general lines that historical development had traversed over the ages. The first stage in the formation of civilizations came when some Aryans, maybe only small numbers of them, subjected people of other races, leading to a long period of cultural flourishing. This subjugation of inferior peoples played an essential role in the development of higher civilizations, according to Hitler. He stated,

> Thus, for the formation of higher cultures the existence of lower human types was one of the most essential preconditions, since they alone were able to compensate for the lack of technical aids without which a higher development is not conceivable.... Hence it is no accident that the first cultures arose in places where the Aryan, in his encounters with lower peoples, subjugated them and bent them to his will.[77]

If these comments presaged Nazi policies of forced labor for those considered racially inferior, his description of their ultimate fate was even more ominous. He compared the exploitation of "lower human beings" in the rise of earlier civilizations with the use of animals, such as horses. They helped the Aryan masters establish a higher culture and create new technologies, ultimately "permitting him to do without these beasts. The saying, 'The Moor has worked off his debt, the Moor can go,' unfortunately has only too deep a meaning." This does not necessarily imply genocide, but it does suggest that the Aryans would get rid of "lower human beings"—one way or another—once they were no longer beneficial to the Aryans.

Though he did not offer any specific historical examples in this particular passage, he was certainly implying that ancient civilizations, such as the Mesopotamian, Egyptian, Greek, and Roman civilizations, were developed by Aryans. Many other Aryan and Nordic racial theorists in early twentieth-century Germany upheld similar views. Elsewhere in *Mein Kampf* he hinted at this, claiming that the Aryans in the north in ancient times had not been able to develop their latent abilities because of the harsh environment, but Greek civilization had flowered because of the hospitable climate. Here Hitler was by no means embracing an environmental determinist position, because he insisted that "the glorious creative ability was given only to the Aryan, whether he bears it dormant within himself or gives it to awakening life, depending whether favorable circumstances permit this or an inhospitable Nature prevents it."[78] The environment, then, was secondary to racial character, as Hitler clearly affirmed, "The inner nature of peoples (*Völker*) is always determining for the manner in which outward influences will be effective."[79] Only the Aryans and no other race could develop advanced cultures in hospitable climates. In a 1927 speech he not only articulated his view that the Aryan race was the sole founder of all higher culture, but he also explained that many lower people are ill-equipped for this racial struggle and thus die out.[80]

One of the clearest expressions of his view that Greek and Roman civilizations grew from Aryan roots came during an August 1920 speech in Munich. Here he stated that Aryans "were in reality the originators of all the later great cultures." The ancient civilizations in Egypt, Persia, and Greece were all founded by blond-haired, blue-eyed Aryans, he claimed.[81] Throughout his life he continued to uphold this view of the Aryans as the driving force behind the history of civilization. In his December 11, 1941, speech, in which he declared war on the United States, he posed as the defender of European civilization against external foes. He rehearsed earlier struggles of the Greeks, Romans, and Germanic peoples against their enemies. He explained, "There was a time when Europe was that Greek island into which Nordic tribes penetrated in order to light the flame for the first time that has since slowly but steadily begun to enlighten the world of man."[82] Thus he clearly ascribed Germanic origins to Greek culture. After the Greeks the cultural torch was passed to the Romans and later to the Germanic peoples, who were the primary representatives of European civilization.

Many of Hitler's associates knew about his love for ancient Greece and Rome as earlier Aryan civilizations. In the midst of a discussion about the constancy of racial characteristics in a 1932 speech, Hitler stated, "I can see the virtues and vices of our German *Volk* in the Roman authors just as clearly as I perceive them today."[83] This clearly indicates that he saw racial continuity between the Romans and present-day Germans. On several occasions during his Table Talks, Hitler referred to Greeks as having Germanic ancestry. "In Greece and Rome," he alleged, "the German spirit could first develop itself!"[84] Close colleagues of Hitler, such as his lawyer Hans Frank and his press chief Otto Dietrich, remembered Hitler's belief that the Greeks were Germanic and their culture was an example of Aryan creativity.[85] Alfred Rosenberg, one of the chief Nazi ideologists, even recalled that Hitler was suspicious of the study of the ancient history of central Europe, since the Germanic tribes there were living in hovels, while the Greeks and Romans were building impressive temples and other buildings.[86] On April 8, 1941, Goebbels recorded in his diary a conversation he had with Hitler that day about antiquity. Hitler refused to allow Athens to be bombed because of his regard for ancient Greek culture. (This was in stark contrast to his attitude toward Eastern Europe, where Hitler recommended destroying all their cultural artifacts and monuments, since they were allegedly inferior.) He then praised Greek and Roman culture, calling the Augustan Age the pinnacle of history.[87]

If the cultural attainments of the Aryans were so important to Hitler, was culture ultimately more important than race? Is Frederic Spotts right to argue that "power was for Hitler ultimately an instrument for achieving his cultural ambitions," and Hitler "saw culture as the supreme value in itself"?[88] Spotts is right to point out that Hitler showed great concern for cultural achievements. As a youth he was an aspiring artist, he patronized the Wagner festival, he dreamed of creating monumental architecture in Berlin, and he planned to make the home city of his youth, Linz, Austria, a cultural capital in Europe. In a 1937 speech, after stressing the importance of the arts in the life of the German people, he stated,

> This state shall neither be a power without culture nor a force
> without beauty. For the armament of a Volk is only morally justi-
> fied when it is the sword and shield of a higher mission. Therefore
> we are not striving for the brute strength of someone like Genghis
> Khan, but instead for an empire of strength which is instrumental
> in shaping a strong social and protected community as the support
> and guard of a higher culture![89]

Cultivation of the arts was a high priority for him, and he did see war and genocide as ways of promoting culture, as Spotts astutely argues. But how do war and genocide advance culture?

As we have seen, Hitler was a biological determinist, believing that culture was the product of hereditary traits. He thought his own preferences for classical forms of art were biologically ingrained in the Aryan psyche. Those with real Aryan instincts, he thought, would prefer the classical forms. Modernist art, on the other hand, was a sign of cultural degeneration rooted in instincts of inferior races or mentally inferior individuals. He often associated modernist art with blacks and Jews. When Hitler opened the House of German Art in Munich, he portrayed modernist art as atavistic, stating, "When we know today that the evolution of millions of years, compressed into a few decades, repeats itself in every individual, then this art, we realize, is not 'modern.' It is on the contrary to the highest degree 'archaic,' far older probably than the Stone Age." This statement shows that Hitler not only believed in human evolution, but he also endorsed Haeckel's recapitulation theory, which claimed that each organism in its embryological development repeats earlier stages of evolutionary history. His use of recapitulation theory to dismiss art as primitive implies that the artists themselves were not as high on the evolutionary scale. Interestingly, in this same speech Hitler discussed the "new human type" that Nazism was creating, a healthier, stronger people. He apparently saw this new humanity as a resurrection of the glories of the classical world, for he added, "Never was humanity in its external appearance and in its frame of mind nearer to the ancient world than it is today."[90] He wanted to purge German art of the newer, modernist forms and return to the healthy instincts of the Greeks and Romans.

While he had a dim view of modernist art, Hitler was enthralled with modern science and technology, which he considered the product of Aryan ingenuity. He pointed to the superiority of European (and especially German) science, while ignoring the many contributions of Jews to modern scientific and medical discoveries. The United States had achieved its awesome technological prowess, he asserted many times, because Germanic blood predominated in its population. Further, he told his colleague Otto Wagener that all great industrial states are Nordic, so he concluded that "the capability of industrial organization is also race-specific."[91] In a 1927 speech he told his audience that all great inventions, including innovations in transportation, have come from Nordic Germans. He stated, "Humanity owes everything great to the struggle and to a race that has triumphed in it." This race, of course, was the Aryan or Nordic race (he used both terms in this speech). "We see before us," he asserted, "that obviously the bearer of all culture and of all humanity is the Aryan."[92] He continued to believe in the exclusively Aryan origins of civilization and advanced culture throughout his career, for he reiterated this point in a private speech to military officers in 1942.[93]

Ultimately, Hitler's view of the relationship between racial characteristics and advanced culture throughout history was somewhat circular.

Figure 3.6   Nazi poster: "Degenerate Music" with ape playing sax

Any advanced civilization—Mesopotamian, Egyptian, Greek, Roman, and so on—must have been founded by Aryans, he reasoned, because only Aryans have the requisite abilities to found such civilizations. The Aryans who remained in northern Europe, living in huts while ancient Greece and Rome flourished, were conveniently excused for their lack of cultural accomplishments, because the climate was too harsh to allow them to exercise their innate abilities. Hitler seemed to know little or nothing about the accomplishments of Chinese or Japanese civilizations, either.

If culture was determined by hereditary traits, then the way to promote culture would be to steer biological evolution to higher levels. In *Mein Kampf* Hitler claimed that a state should not be judged by its cultural attainments, but by its ability to preserve the race, which was the basis for all achievements.[94] Since Hitler considered Germans biologically and culturally superior to other races, he thought that whatever means increased the numbers and quality of Germans on the earth would lead to higher cultural achievements. Warfare and extermination of other races to make room for the higher Aryan race was thus the precondition for the development of higher culture. Warfare, though it might be inimical to the development of culture in the short term, was necessary to establish the conditions for the development of culture, Hitler explained in *Mein Kampf.* The Persian Wars laid the groundwork for cultural flowering of the Age of Pericles, and the Punic Wars made Roman culture possible. Thus Hitler justified his own focus on military power as a necessary means to promote a later cultural renaissance.[95]

Hitler articulated essentially the same point in the chapter on "Race, Struggle, and Power" in his *Second Book*, where he insisted that the value of one *Volk* is not the same as the value of another. He called some races valuable and others worthless. While he claimed that the value of a *Volk* is based on its biological properties, he admitted that the only way to properly judge the value of a *Volk* was by evaluating its cultural achievements. "The ultimate expression of this overall valuation" of a *Volk*, he stated, "is the historical cultural image of a people [*Volk*], in which the sum of all the rays of its genetic qualities—or the racial qualities united in it—are reflected." He then admonished his fellow Germans to remain proud of their racial value and the concomitant cultural attainments. He then even more explicitly linked the cultural level of a people with its racial character, stating, "When I speak about the inner racial value of a people, I assess this value based on the sum total of the people's visible achievements, thus acknowledging at the same time the presence of the particular personal qualities that represent the racial value of a people and create its cultural image." Culture is thus the outward expression of inner racial characteristics that are expressed by creative individuals.[96]

Racism and Morality

Since he considered the Aryan race superior to all other races, and since they were locked in an ineluctable struggle for existence with other races, whatever benefited the Aryan race in that struggle was morally justified and even morally praiseworthy. He considered it nothing but pacifistic twaddle to moralize about the natural struggle. In *Mein Kampf* he proclaimed, "When the nations (*Völker*) on this planet struggle for their existence—when the question of destiny, 'to be or not to be,' cries out for a solution—then all considerations of humanitarianism or aesthetics crumble into nothingness."[97] In a secret speech in 1937 to future Nazi leaders, he remarked, "Today the knowledge of the significance of blood and the race exalts itself above a humane worldview."[98] Hitler consciously rejected humanitarian concerns, especially if they hindered the German *Volk* from defeating its racial enemies in the struggle for existence.

Hitler frequently insinuated that the struggle between races was necessarily brutal. No feelings of sympathy or pity should cloud one's judgment. In a 1920 speech he told a cheering Munich audience that he was not going to allow those of other races—and here he was referring to the Jews—to cross him. "If, however, a large race consistently destroys the conditions of life of my race, I do not say, that it does not matter to me where he belongs. In that case I say that I belong to those who, when they receive a blow on the left cheek, pay back with two or three."[99] Hitler obviously did not think it was appropriate to apply the morality of the Sermon on the Mount to the racial struggle. As in this example, Hitler often exulted in the violence of the racial struggle and disparaged those wanting peaceful racial relations.

In a 1929 speech he not only explained the importance of a violent struggle, but he also linked it to cultural improvement. The first element in the Nazi worldview, he explained, is the principle that the strong always triumph and those with the greatest value always shape the world. Humans are selected based on their power. Hitler then continued, "*The domination of the white race is not the product of the mutual understanding of peoples, but they have slowly raised themselves by bloody struggle, and have given the world what we call culture. Every achievement in the world has arisen through struggle.*"[100] While acknowledging that the racial struggle is bloody, he insisted that it produced beneficial results. On the other hand, the "mutual understanding of peoples," by which Hitler meant international agreements, were counterproductive.

In fact, Hitler believed that Aryans had an inherent feeling or instinct to dominate other races. In 1932 he told industrial leaders that since the dawn of antiquity the white races had taken a position of leadership in the world. The past several centuries had proven this, as the white race had expanded at the expense of other races. Hitler then gave three examples of how the white race had used brutal means to conquer and subdue indigenous peoples: the British in India, the Spanish in Latin America, and the conquest of North America, which did not occur through democratic means or according to international law, he explained, but from "a consciousness of what is right which had its sole roots in the conviction of the superiority and thus the right of the white race." He applauded the white race for subduing other races by force and warned that if they abandoned this expansionist sentiment, they would become overpopulated, like the Chinese. This feeling of dominance was not only the basis of colonization, but it had allowed the Aryan or Nordic race to establish the German state, too. The drive for superiority might be camouflaged at times, but always "it was the exercise of an extraordinarily brutal right to dominate."[101] Brutality was morally justified, then, if it advanced the welfare of the allegedly superior race.

Hitler's highest priority in life was to improve the human species, to advance evolution. Helping Aryans win the struggle for existence against other races was crucial to achieving his vision. Morality itself was measured by whether or not it benefitted the German people in their struggle. In the opening passage of a chapter of his *Second Book* entitled "Struggle, Not the Economy, Secures Life," he articulated the relationship between morality and struggle: "Therefore ideals are healthy and appropriate as long as they help to reinforce a *Volk*'s inner and collective strength, so that these forces can contribute in carrying out the struggle for life. Ideals that do not serve that purpose, even if they appear a thousand times beautiful outwardly, are nevertheless evil, because they gradually distance a *Volk* from the reality of life."[102]

In a private speech to military leaders a few months after the outbreak of World War II, Hitler explained the reason for the war: "Someone might reproach me: Struggle and yet again struggle. I see in struggle the destiny of all beings. No one can escape the struggle, as long as he does not want to be defeated." He then specifically called the present war a "racial struggle." This racial struggle was caused by the need of Germany's growing population for more living space (*Lebensraum*).[103] This evolutionary view of the racial struggle for space underpinned Hitler's quest for military expansion, as we shall see in chapter 8.

CHAPTER FOUR

Morally Upright Aryans and Immoral Jews

Racial Struggle between Aryans and Jews

In one of the most famous passages of *Mein Kampf*, Hitler declared, "Hence today I believe that I am acting in accordance with the will of the Almighty Creator: by defending myself against the Jews, I am fighting for the work of the Lord." The Jews obviously played a central role in Hitler's thinking about the racial struggle for existence. They were racial enemy number one. However, what was this "work of the Lord" that Hitler thought he was advancing? If we examine the context of his remark, we can see that Hitler believed that the Jews were a threat to the very existence of humanity. Just prior to this famous quotation, Hitler explained how dangerous the Jews were:

> The Jewish doctrine of Marxism rejects the aristocratic principle of Nature and replaces the eternal privilege of power and strength by the mass of numbers and their dead weight. Thus it denies the value of personality in man, contests the significance of nationality and race, and thereby withdraws from humanity the premise of its existence and its culture. As a foundation of the universe, this doctrine would bring about the end of any order intellectually conceivable to man. And as, in this greatest of all recognizable organisms, the result of an application of such a law could only be chaos, on earth it could only be destruction for the inhabitants of this planet. If, with the help of the Marxist creed, the Jew is victorious over the other peoples of the world, his crown will be the funeral wreath of humanity and this planet will, as it did millions of years ago, move through the ether devoid of man.[1]

The "aristocratic principle of Nature" was a phrase that was derived from Haeckel, who used the term frequently to emphasize the inegalitarian social implications of evolutionary theory.[2] By resisting the "aristocratic principle of Nature," Jews were more than just a threat to future evolutionary progress. They would completely demolish the highest organism that nature had produced over eons of time, and all advanced culture would vanish. Because the stakes were so high—the very existence of humanity—he considered any means to outstrip the Jews in this competition-to-the-death morally justified. Defeating the Jews in the racial struggle for existence and replacing them with the higher Aryans was essential, so that humanity could continue its upward evolution, in Hitler's view.

As with his other ideas, Hitler was by no means original in construing the contest between Aryans and Jews as a Darwinian struggle for existence. One of Haeckel's students, Willibald Hentschel, became a prominent anti-Semitic publicist in the early twentieth century. He opened his major book on Aryan racial theory, *Varuna* (1901), by explaining that his racial views were scientific, since they were based on Darwinism.[3] His publisher, Theodor Fritsch, one of the most infamous anti-Semitic publicists in the early twentieth century, boasted that Hentschel had placed anti-Semitism on a scientific foundation. Fritsch even claimed that *Varuna* "counts as the programmatic statement of 'Hammer,'" his popular anti-Semitic journal.[4] Houston Stewart Chamberlain also construed the competition between the Teutons and Jews as a racial struggle for existence. Even though he rejected some aspects of Darwinian theory, he remarked that Darwin had correctly hit upon the ideas of racial struggle and selection, which Chamberlain incorporated into his anti-Semitic racial thought.[5]

Hitler saw the racial struggle against the Jews as apocalyptic. He sincerely believed in a Jewish world conspiracy that involved both Jewish capitalists and Jewish communists. He was completely duped by the forged *Protocols of the Elders of Zion*. He considered his fight against the Jews an act of defense against the Jewish conspiracy.[6] That is why he encouraged his fellow Germans in January 1923 to focus on the racial struggle instead of the class struggle. They must take vengeance on "those who push them into the abyss."[7] Hitler thought the racial struggle would result in one of two outcomes for the Aryan race: complete victory or total annihilation. In a 1922 speech he stated that there were "only two possibilities in the incredibly great fight: Either victory for the Aryan side or its destruction and victory for the Jew."[8] Hitler was always confident that the Aryans would triumph, since they were superior racially to their enemies. Even as his armies retreated in the latter stages of World War II, he thought that just a little more willpower on their part would tip the balance in their favor. Only as the Soviets closed in on his Berlin bunker in April 1945 did he despair. Instead of blaming his own misguided policies, he blamed the German people and declared that they deserved their ignominious fate. Nonetheless, even while consigning them to destruction, he encouraged them to continue his racial policies so they could regain their strength in the future. Thus, even in these dire circumstances, he still held out hope for a renascence of German power in the future.

Because Hitler believed that the Jewish quest for world dominion "lies profoundly rooted in their essential character," the only way to defeat it was to get rid of the Jews, one way or another. All Jews were implicated in this world conspiracy, since the Jews are a race "which today more than ever is conscious of a mission to impose its bloody oppression on the whole world." Their conniving could lead Germany "into the abyss," so he saw anti-Jewish measures as a morally necessary defense against Jewish depredations. This struggle against the Jews had a moral dimension, because the Jew "goes his way, the way of sneaking in among the nations and boring from within, and he fights with his weapons, with lies and slander, poison and corruption, intensifying the struggle to the point of bloodily exterminating his hated foes."[9] Thus, in Hitler's twisted view the struggle against the Jews was a struggle against immorality and racial extermination. Hitler's perspective was, to use Saul Friedländer's apt phrase, "redemptive anti-Semitism."[10] Indeed Hitler construed this struggle against Jewish immorality as part of the Darwinian struggle for existence. In his *Second Book* he stated that in the struggle for existence the Jew uses immoral tactics, exercising "shrewdness, cleverness, cunning, disguise, and so on, which are rooted in the character of his people. They are stratagems in his fight to preserve life, just like the stratagems of other peoples in military conflict."[11] To win the struggle for existence Germans needed to counter Jewish immorality.

However, defeating the Jews had yet another moral dimension. As a biological determinist, Hitler believed that moral character was not shaped primarily by upbringing and education. Rather, he thought hereditary traits were the most important factor determining one's behavior and character. This is important to grasp, because many scholars have ignored this equation of biology and behavior that is crucial for understanding Hitler's anti-Semitism. For instance, Jeffrey Herf argues forcefully in his recent book, "Hitler and his associates decided to murder the Jews in Europe because of what they believed 'international Jewry' *did*, far more than because of the way Jews were said to look. From the Nazis' perspective, it was the Jews' actions, not their bodies, that justified mass murder."[12] Indeed, Herf does prove that Nazi propaganda blamed the Jews for immoral deeds, and he is correct that Nazis were not persecuting Jews for their physical characteristics. However, this misses an important point. As the burgeoning literature on Nazi eugenics has shown—and as I demonstrate below—Hitler (and many other Nazis) did not draw such a dichotomy between biology and behavior. What the Jews did, Hitler thought, was a product of their heredity. Yes, Hitler and the Nazis painted the Jews as criminals, but they thought criminality was rooted in their biological fabric.[13] In 1943 a Nazi directive to the German press declared, "Jews are criminals....Jewry as a whole springs from criminal roots and is criminal by disposition. The Jews are not a nation like other nations, but bearers of hereditary criminality."[14]

Since Hitler defined races as biological units possessing similar hereditary traits, each race manifested different moral traits. In a 1932 speech

Figure 4.1　Article from Nazi periodical: "The Criminal Jew"; captions by photos indicate the crimes each committed: swindler, counterfeiter, pickpocket, etc.

to industrialists he explained that one of the key factors influencing politics is the "inner value of a people (*Volk*), which is passed on again and again through the generations as hereditary factors and hereditary qualities." He then explained that these hereditary qualities include moral characteristics: "It is certain that definite character traits, definite virtues and definite vices always reappear, so long as their inner nature, the composition of their blood, has not altered essentially." Virtues and vices, then, were fixed hereditary traits that differed from person to person and from race to race. This view of hereditary moral characteristics gave Hitler hope for the resurrection of the German people, since they still had inherent biological vitality that would eventually overcome their adverse temporary circumstances. Defeat in World War I and the humiliation of Versailles were only minor bumps in the historical road that could not obliterate the superior biological quality of the German people.[15]

Hitler's emphasis on hereditary continuity from generation to generation and unchanging racial features seems at first glance inconsistent with biological evolution. However, this is not really the case. Hitler recognized that evolution is such an incredibly slow process that in the limited time of known human history—only several thousand years—little biological change could occur. These moral traits had evolved over vast stretches of time, so they could not be quickly altered. For all practical purposes, when dealing with centuries or even millennia rather than eons, races had fixed essences. Thus, even though he believed that races had formed through evolutionary processes, racial characteristics could only change over extremely long periods of time. Therefore, even when he seemed to imply that races had unchanging essences or hereditary character that was unalterable, he did not mean to imply that they had not or could not change over geological time.

Moral Superiority of Aryans

Not only did Hitler think that Aryans were physically and intellectually superior to people of other races, but he also considered them morally superior. In *Mein Kampf* he even argued that the moral superiority of the Aryans—not their intellectual powers—was the main source of their greatness and their ability to create higher culture and civilization. After discussing how all organisms have an instinct of self-preservation, he explained that in humans this instinct broadens to include concern for the preservation of social groups, beginning with families and expanding outward from there. He believed that Aryans have the greatest measure of concern for their community, which has given them the unique ability to establish advanced cultures and civilizations:

> The self-sacrificing will to give one's personal labor and if neces-sary one's own life for others is most strongly developed in the Aryan. The Aryan is not greatest in his mental qualities as such, but in the extent of his willingness to put all his abilities in the service of the community. In him the instinct of self-preservation has reached the noblest form, since he willingly subordinates his own ego to the life of the community and, if the hour demands, even sacrifices it.[16]

Right after discussing the Aryan's inner instinct for community-building, Hitler linked this inherent idealism to the evolutionary process:

> Here the instinct of knowledge unconsciously obeys the deeper necessity of the preservation of the species, if necessary at the cost of the individual, and protests against the visions of the pacifist windbag who in reality is nothing but a cowardly, though cam-ouflaged, egoist, transgressing the laws of evolution; for evolution requires willingness on the part of the individual to sacrifice him-self for the community, and not the sickly imaginings of cowardly know-it-alls and critics of Nature.[17]

For Hitler the essence of Aryan morality was the inner inclination or instinct to sacrifice one's own individual existence for the life of the community, which would preserve the human species and advance human evolution.

Many historians have analyzed Hitler's anti-individualism, his disdain for individual rights, and his insistence that the nation or race takes pre-cedence over individual liberties.[18] He expressed these ideas frequently. For instance, in a 1940 speech to military officers he expounded on why they were laying their lives on the line in the war. "It is of no import whether the individual among us lives—what must live is our *Volk*."[19] An individual perishing in warfare—or in the concentration camps—was of no significance for Hitler, as long as it made Germany stronger on the whole. Hitler's complete lack of concern for individual life resulted in policies that squashed civil liberties and destroyed individual freedoms.

However, no historians have analyzed how these ideas about the relationship between the individual and the collective radiated out from an evolutionary ethic.[20] Since Hitler believed evolutionary progress was the highest good, he subjected the interests and rights of the individual to that of the collective. In the evolutionary process, myriads of individuals in every species perish, after all, for the benefit of the species. The individual has little or no significance in the wider evolutionary scheme. Hitler believed that this subjection of the individual to the race promoted a higher good, and he encouraged Aryans to willingly sacrifice themselves on behalf of their fellow Aryans. However, if they were not inclined to such altruistic behavior, he was willing to help them out and sacrifice them on the altars of evolutionary progress.

Indeed Hitler overtly linked his anti-individualism with the evolutionary process in nature. He articulated this idea many times, both publicly and privately. In a private conversation in January 1941, Hitler stressed the need to study the laws of nature, so that one does not kick against the goads. He continued by stating, "If I want to believe in a divine command, it can only be this one: to preserve the species! One really should not value the life of individuals so highly."[21] This statement suggests that Hitler derived the highest moral precept from nature, and that command was to preserve the human species. Hitler clearly stated in *Mein Kampf* that the highest purpose of human existence is not to preserve the state, but rather to preserve the species. The state's sole purpose, he thought, is to contribute to the biological maintenance of humanity.[22]

Hitler was sure that his disdain for the individual was consistent with the laws of nature. After all, nature sacrificed multitudes in the struggle for existence to make way for the fittest to triumph. He scoffed at humanitarian concern for the individual, because by "sparing individuals, the future of millions is sacrificed."[23] He reiterated this in a 1928 speech, stating, "Nature is pitiless with the individual, but full of pity for the aggregate."[24] Hitler's anti-individualism, then, was modeled after the Darwinian struggle for existence. It reflected his overarching concern for the future evolutionary development of humanity. His concern for humanity was completely abstract, of course, as it had to be, since individuals counted for little or nothing.

Though Hitler clearly rejected the liberal stress on the value of the individual, he did not embrace the interests of all humanity, despite his rhetoric about helping the human species. As we have already seen, whenever Hitler outlined the National Socialist worldview, he always stressed the importance of race. However, Hitler sincerely believed that advancing the cause of the Aryan race would benefit the human species, since he considered Aryans the most highly evolved form of humanity, who had the best prospects for surviving and perpetuating the species. In October 1933 he explained that National Socialism does not put the individual or humanity at the center of its attention, but rather the German *Volk*: "The lone individual is short-lived, the *Volk* is lasting. While the liberal world outlook by according the individual a god-like status, must of necessity lead to the destruction of the *Volk*, National Socialism wishes to preserve the *Volk* as such, if necessary at the expense of the individual."[25] Earlier he had remarked in *Mein Kampf* that public policy relating to sexuality "must be determined by concern for the preservation of the health of our people in body and soul. The right of personal freedom recedes before the duty to preserve the race."[26]

In any conflict of interests, the health and welfare of the race always trumped the well-being of the individual. If racial interests were at stake, one did not even have a right over one's own body, according to Hitler: "We must also do away with the conception that the treatment of the body is the affair of every individual. There is no freedom to sin at the cost of posterity and hence of the race."[27] Of course, once they were firmly in power, Hitler and his fellow Nazi leaders held in their hands the power to decide when the interests of the race were at stake, allowing them to persecute anyone disagreeing with Nazi ideology or policy.

In Hitler's view, then, the Aryan race was the most advanced race primarily because it was the most altruistic. True Aryans sacrificed their own interests for the interests of others—at least if the others are fellow Aryans. They had an instinctive inclination to follow the Nazi motto, "Common good before individual interests." This spirit of self-sacrifice and concern for the community reflected the "moral and ethical sense of the Germanic race," according to the Nazi Twenty-Five Point Program.[28]

Another instinctive moral trait of Aryans making them superior to other races, in Hitler's view, was their penchant for diligence. (This is ironic, of course, since Hitler had very little of this allegedly Aryan trait.) In his early speeches Hitler often emphasized the Aryan's positive attitude toward labor. According to a contemporary report of a

1920 speech, "Hitler sees the chief difference in the character of the Germanic-Aryan and the Jewish race in their fundamentally different positions toward work as an end in itself or as a means to an end." Seeing labor as an end in itself was, of course, the position of Aryans, while work as a means to an end characterized the Jews.[29] Earlier in the same year he explained in even greater depth how the Aryans came to be such a diligent people. Their struggle against the harsh Nordic elements made a work ethic absolutely necessary for survival. Thus, over long ages the Aryans had developed a hereditary love for labor. In this speech Hitler asserted, "Aryanism means the moral conception of labor and through it what we speak about so often today: socialism, a sense of community, [and] common good before individual interests."[30] Aryans, then, represented the pinnacle of moral character, because they were both diligent and altruistic.

At this point Hitler was caught in a contradictory position (though he was not alone in this, for many proponents of evolutionary ethics before him embraced the same contradiction). He seemed to treat some forms of moral character—diligence, self-sacrifice, loyalty, and so on—as objectively higher or better. Here Hitler was mirroring the values and moral presuppositions of his society, for these moral ideals permeated German society. However, Hitler's evolutionary ethic effectively undermined any fixed moral principles, for they were only valid as long as they contributed to winning the struggle for existence or promoting human evolution. Thus, despite thinking that Aryans have a fixed, instinctive moral character that made them higher than all other people, Hitler did not need to adhere to that morality if nature—or at least what Hitler took to be nature's will—dictated otherwise. After all, selfishness and egoism, diligence and laziness, are all the same to nature.

Hitler's insistence that moral traits were primarily hereditary did not mean that he completely dismissed the efficacy of education in shaping moral character. Education could help reinforce healthy instincts. He lamented the lack of moral instruction in schools and wanted to see them instill loyalty, the spirit of sacrifice, and discretion in German children.[31] He also claimed that the army played a vital role in inculcating morality in young German men. "The army trained men in idealism and devotion to the fatherland and its greatness while everywhere else greed and materialism had spread abroad," he declared.[32] Nonetheless, again and again Hitler intoned that moral education was only efficacious for those with hereditary traits disposing them to moral behavior.

Jews as Biologically Immoral

If the Aryan was the paragon of moral virtue in Hitler's worldview, the Jew was the epitome of evil. Hitler depicted Jews as the embodiment of every sinful and shameful behavior and attitude imaginable. They were the very emblem of evil, he declared: "In his vileness he becomes so gigantic that no one need be surprised if among our people the personification of the devil as the symbol of all evil assumes the living shape of the Jew."[33] Hitler imbibed and perpetuated all the centuries-old negative stereotypes of Jews as immoral, portraying them as greedy, deceitful, and sexually perverted. In *Mein Kampf* he filled page after page detailing the alleged immorality of the Jews, whom he blamed for undermining German moral fiber by flaunting sexuality in the press and the fine arts. He also accused them of sponsoring prostitution, and he repeatedly called them liars. "Was there any form of filth or profligacy, particularly in cultural life, without at least one Jew involved in it?" he queried.[34]

In Hitler's ideology, immorality is biologically determined, just as altruism is, so the Jews' vices could not be remedied by a good upbringing or education. In a 1923 interview with an American journalist, Hitler clearly explained his view that crime is hereditary. In the midst of a discussion about the need to hinder the propagation of hereditary problems, he stated that the hereditarily ill need to be isolated to keep them from reproducing. This applied to those with hereditary moral problems, as well as those with congenital physical or mental illnesses. He stated, "I would isolate the criminal as well as the person suffering from some physical taint. One disease breeds many. One pimp makes ten. One criminal in the course of a few generations, infects hundreds with the seed of crime, insanity and disease."[35] In a 1934 speech he divulged this viewpoint again by accusing the communists of unleashing the "criminal instincts" of subhumanity.[36] Hitler thought that not only crime, but all immoral character passes on from parents to their progeny, just as physical traits do. Since he thought the evil moral character of Jews was hereditary, he opposed any attempts to try to assimilate Jews to German society and culture. They would remain immoral, no matter how much they tried to assimilate.[37]

In most of the passages we have already examined about the upstanding moral qualities of the Aryans, Hitler contrasted them with the nefarious Jews. Immediately after discussing the self-sacrifice and idealism of Aryans in *Mein Kampf*, Hitler turned his attention to the Jews, depicting them as the exact moral opposite of the Aryans: "In the

Jewish people the will to self-sacrifice does not go beyond the individual's naked instinct of self-preservation." Even though they seem to have a sense of communal spirit, this is only a herd instinct that quickly dissolves when they are no longer in danger:

> His sense of sacrifice is only apparent. It exists only as long as the existence of the individual makes it absolutely necessary... The Jew is only united when a common danger forces him to be or a common booty entices him; if these two grounds are lacking, the qualities of the crassest egoism come into their own, and in the twinkling of an eye the united people turns into a horde of rats, fighting bloodily among themselves.

Hitler continued this rant about the Jews by claiming that they are so consumed with selfishness that if they were the only race existing, "they would try to get ahead of one another in hate-filled struggle and exterminate one another."[38] Not only in this passage, but every time Hitler bashed Jews for their selfish, immoral behavior, he assumed that it was an inherent biological trait that would persist as long as Jews continued reproducing.

Hitler's concern about Jewish immorality remained entrenched in his ideology to the end of his life. In 1937 he told the crowds gathered for the Nazi Party Rally at Nuremberg that the Jewish race "is neither spiritually nor morally superior, but in both cases inferior through and through. For unscrupulousness and irresponsibility can never be equated with a truly brilliant disposition."[39] Lack of scruples was thus part of the Jewish racial character that made them an inferior race. These attitudes shaped Hitler's policy toward the Jews to the end of his regime.

Sexual immorality was another trait Hitler frequently associated with Jews. When recounting in *Mein Kampf* his own (probably fictionalized) path to embracing anti-Semitism in Vienna, he stressed the role of the Jews in promoting sexual immorality in Viennese culture and especially their alleged role in supporting prostitution. "When thus for the first time I recognized the Jew as the cold-hearted, shameless, and calculating director of this revolting vice traffic [prostitution] in the scum of the big city, a cold shudder ran down my back."[40] He devoted an entire 1921 speech to "The German Woman and the Jew," wherein he contrasted the noble love that German men have for women with the carnal indulgence of Jews. Jews have no capacity for love, Hitler claimed, but only engage in sex to fulfill their fleshly appetites. Their different moral dispositions shape their attitudes toward women: "The *German* can devote and sacrifice his life for his *love* and his *woman*, but for a woman the *Jew* can only—pay!"[41]

Jewish sexual immorality was a grave concern for Hitler for three main reasons. First, through the press, theater, and film they were allegedly corrupting the pure instinctive morals of Germans. Second, as we will see later, Hitler opposed racial mixing as a danger to the hereditary health of the German people. He thought that Jews were a menace to morally pure but sometimes naïve German girls and women, whom they seduced. The resulting "bastardized" offspring was below the physical and moral level of pure Aryans. Third, Hitler thought that Jews were using racial mixture consciously as a means to destroy Germany. Their sexual liaisons with German women were thus part of the worldwide conspiracy that Hitler thought Jews were orchestrating.

In a famous passage in *Mein Kampf* Hitler accused Jews of using sex to undermine the German people:

> With satanic joy in his face, the black-haired Jewish youth lurks in wait for the unsuspecting girl whom he defiles with his blood, thus stealing her from her people. With every means he tries to destroy the racial foundations of the people he has set out to subjugate. Just as he himself systematically ruins women and girls, he does not shrink back from pulling down the blood barriers for others, even on a large scale. It was and it is Jews who bring the Negroes into the Rhineland, always with the same secret thought and clear aim of ruining the white race by the necessarily resulting bastardization, throwing it down from its cultural and political height, and himself rising to be its master.[42]

It is not clear here if Hitler meant that the "satanic joy" of the Jewish youth was anticipation for the carnal pleasures of sexual relations, or if it was anticipation of undermining his racial enemies by this sexual subterfuge. Likely it was both. In any case, Hitler portrayed it as a systematic and conscious effort to destroy or subdue their foes in the racial struggle for existence.

In this passage Hitler also accused the Jews of using other races to undermine Germans. Black Africans serving in the French military were part of the occupation forces in the Rhineland. The Nazis and other radical nationalists, outraged at the children fathered by these African troops, called them "Rhineland bastards." In a June 1922 speech Hitler blamed the Jews for relying on other races to subdue Germany, stating,

"And as cultural guardians of Jewish capitalism Chinese executioners' assistants stand in Moscow and black ones in the Rhineland."[43] Hitler often associated the Jews with other allegedly inferior races, whom they enlisted to defeat the noble Aryans.

The hereditary trait that Hitler most often associated with the Jews was an inclination to avoid work. In a 1922 speech castigating the Jews, he stated,

> The Aryan understands labor as the foundation for the preservation of the people's community (*Volksgemeinschaft*), the Jew sees it as the means to exploit other peoples (*Völker*)....It does not matter if this individual Jew is "decent" or not. He bears within himself the character traits that nature has granted him, and he can never free himself from it.[44]

Jews, then, were by nature work-shy and exploited others in order to survive, he claimed. In a 1920 speech he had made similar claims, arguing that the Jews' "egoistical conception of labor and thus mammonism and materialism" were hereditary traits inherent in Jewish blood.[45] They did not make any real economic contributions to society. On the contrary, they used clever, sly tactics to force others to produce for them. They epitomized the greedy capitalist exploiting the labor of common men and women, so they would not have to work themselves. They were responsible for everything that Hitler considered oppressive about the German economic system.

In *Mein Kampf* Hitler presented his vision of the Jews' contemporary economic position by outlining the stages of historical development of Jews within European societies. First, Jews began as merchants in a foreign society. Then, because of his "thousand-year-old mercantile dexterity he is far superior to the still helpless, and above all boundlessly honest, Aryans," and therefore he monopolized commerce. The Jews began lending money and introduced interest to their host society. These measures eventually aroused resistance from the host peoples, especially when the Jews reduced land to the status of a commodity. The people then denied Jews the right to own land, but the Jewish "blood-sucker" remained viable by flattering and bribing princes. Later the Jews adopted another subterfuge by trying to pass as Germans; at this point they posed as liberals and benefactors of mankind. Then the Jews took control of the stock market and came to control the capitalist system. At the same time, the Jews wooed the alienated workers, manipulating them in a struggle ostensibly against the Jewish capitalists' own interests. However, "the great master of lies understands as always how to make himself appear to be the pure one and to load the blame on others." The masses gullibly swallowed the Marxist lie that the Jews concocted.[46]

Hitler's vision of the oppressive role of Jews in the economy was based on the idea that Jews were biologically predisposed to mercantile activity, a notion that many prominent social Darwinists before Hitler had upheld, including Büchner and Lenz. For this reason, Hitler constantly attacked them as parasites on the nation. They were not productive, but only lived off the productivity of others. Hitler told an Augsburg audience in May 1923, "Everywhere, in all peoples the Jew snuck in as a deadly parasite, in order to live there from the labor of the host people."[47] They relied on trade, banking, interest, and unearned income to skim off the wealth produced by the German people. Countering a widespread view of the Jews as nomads, Hitler remarked that this was not so, since even nomads do productive work. Jews do not have their own homeland and they wander from place to place not because they exhaust the resources of one place, but because they are expelled by their host, or because they are forced to move after their host dies. Unlike nomads, they are unable to support themselves, but must live on others. They are "like a noxious bacillus" that infects their host society and ultimately destroys it.[48]

Jews cannot form their own states, Hitler claimed, because they lacked the requisite "heroic virtues" and self-sacrifice necessary to build a strong community and preserve a state. Their "egoism of shop-keepers" kept them too weak and disunited to maintain their own soil. Further, their parasitical existence tends to foster "lying hypocrisy and malignant cruelty."[49] According to Hitler, Jews were simply following their instincts when they pursued their parasitic existence. Hitler asserted, "Never yet has a state been founded by peaceful economic means, but always and exclusively by the instincts of preservation of the species regardless whether these are found in the province of heroic virtue or of cunning craftiness; the one results in Aryan states based on work and culture, the other in Jewish colonies of parasites."[50] The Jews' "cunning craftiness" and parasitical practices were simply manifestations of their internal drive to live and reproduce.

Hitler's view of the Jews as a biological or racial entity rather than a religious community, together with his insistence that their biological nature was inherently evil, made a dangerous combination. For Hitler (and other Germans sharing his ideology) the only way to rid the world of immorality was to stop the Jews from passing on their evil hereditary traits. Various methods could lead to Hitler's goal. One plan Nazi officials discussed was deportation of the Jews from Europe. Ultimately Nazis attempted to systematically exterminate all the Jews within their grasp. Sterilization would have been another possibility, though I do not know of any evidence that Nazis seriously considered sterilizing Jews en masse. In any case, while gassing Jews in concentration camps was not an inevitable result of Hitler's evolutionary ethic, nevertheless, ultimately his worldview did contain genocidal tendencies, since it aimed at the eventual elimination of all Jewish hereditary traits from the world.

CHAPTER FIVE

Hitler's Socialism:
Building the People's Community

The Winter Relief Drive

On October 5, 1937, more than four years after coming to power, Hitler bragged that his regime had introduced the "greatest social achievement of all time." What was this program that in the first few years of Nazi rule had "constitute[d] glorious chapters in our *Volksgemeinschaft* [People's Community]"?[1] In this case Hitler was not boasting about cutting unemployment during the worldwide depression, nor was he referring to the Hitler Youth or the German Labor Front, which were assuredly important Nazi programs intended to build German unity. No, it may be surprising, but the prize for the "greatest social achievement of all time" went to the Winter Relief Drive, an annual campaign to raise money for the indigent. Though we should take Hitler's superlatives with a grain of salt, we have abundant evidence that he did highly value the Winter Relief Drive as a vital tool to build and strengthen the German People's Community.

The first annual Winter Relief Drive began in September 1933, when millions of Germans were still unemployed during the Great Depression. According to Nazi statistics, this government-sponsored effort to collect funds for the poor garnered 1,490,760,834 Deutschmarks in the first four years.[2] Giving was supposedly voluntary, but pressure, coercion, and even threats regularly accompanied the pleas for money, so few dared refuse.

This demonstration of concern for the poor gave the Nazis some credibility in their claim to moral leadership. What could be more selfless and altruistic than giving one's resources to help those less fortunate? The Winter Relief Drive was so important to Hitler that up to 1942, even while World War II was raging, he annually delivered a major speech in September or October initiating the campaign. After the winter of 1942–1943, because of the catastrophic military defeats, he retreated from most public speaking, but he still issued a written appeal for the Winter Relief Drive in the fall of 1943. Often he publicly thrust money into the outstretched tins of young Nazis collecting for the drive, demonstrating to the German people his own willingness to sacrifice for the sake of the *Volk*. Of course, publicizing his donations had great propaganda effect, not only making him look good and moral, but also stimulating his compatriots to similar largesse.

However, just as in all other areas of Nazi morality, we need to inquire about the underlying ethic behind the Winter Relief Drive. Why did Hitler and the Nazis want to help the poor, and how did this fit into their broader worldview? Although Hitler did appeal to his fellow Germans' sense of compassion for the plight of the poor, in all of his speeches launching the Winter Relief Drive his greatest emphasis was not on the blessings of helping individuals overcome their adverse circumstances. Rather he stressed the need for building national unity, so Germany could recover its strength on the world stage.

He believed that forging the People's Community through this kind of practical socialism would bring Germany to a position of greater power. In the early years of the Nazi regime he had to mask this somewhat, duplicitously promising that German strength was not a threat to other nations. Flaunting national power would arouse foreign suspicions, which could hinder his long-range foreign policy plans. Nonetheless, before World War II he often stressed the need for unity, and during World War II he was more explicit about the reasons for it. While initiating the Winter Relief Drive during the first year of World War II, he explained that the war would work together with the Winter Relief Drive to unify the German people: "The wartime winter now facing us will find us ever the more prepared to make the sacrifices necessary to ease the struggle for existence for our *Volk*." After exhorting his fellow Germans to greater sacrifice for the war effort, he stated that "the wartime Winter Relief Drive must contribute to making this German *Volksgemeinschaft* [People's Community] stronger than ever before!"[3] In September 1940 he again urged the German people to contribute to the Winter Relief Drive, demonstrating that they are "willing to make any sacrifice this struggle for existence, for our future, will impose upon us."[4] Ultimately, then, the Winter Relief

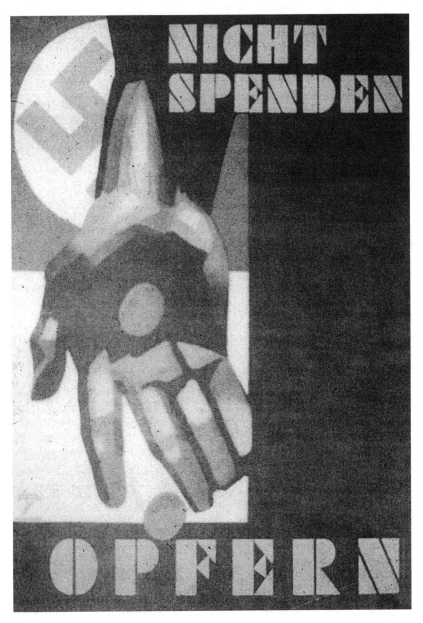

Figure 5.1 Nazi poster: "Don't contribute/[instead] sacrifice."

Drive was an instrument to help the German people win the struggle for existence by making Germany powerful militarily.

Building unity in the People's Community was a goal of the Winter Relief Drive from the very start. It was not an idea that first surfaced during the war to garner greater contributions. Indeed, building national unity by helping the poor seemed like a very noble project, especially since most Germans did not recognize Hitler's expansionist motives linked to it. In his 1933 speech inaugurating this charitable program, Hitler proclaimed that the chief purpose of the campaign was to forge unity among the German *Volk*. He called on his fellow Germans to replace Marxist internationalism with national solidarity. Apparently one purpose of the Winter Relief Drive was to woo the poor away from communist radicalism and build loyalty to the German nation and the Nazi Party. Hitler construed this national unity in racial terms by explaining that the only reasonable kind of unity was "solidarity eternally rooted in the blood." Encouraging his fellow Germans to sacrifice their own individual interests to the interests of the People's Community, he enjoined those Germans with ample means to give liberally to those less fortunate. Winter Relief was thus a means of advancing his nationalist agenda. It would make Germany stronger by promoting internal harmony and peace.[5]

Hitler's views about national unity and the People's Community were forged in an ideological milieu impregnated with social Darwinist ideology, as Peter Walkenhorst has recently demonstrated in his book on radical nationalism in Germany. Walkenhorst portrays social Darwinism and Gobineau's racism as the two main currents shaping radical nationalism in early twentieth-century Germany. He argues that social Darwinism helped mediate a shift toward a new ethic that replaced humanity with the *Volksgemeinschaft* (defined racially). This view was prominent in Pan-German circles around the time of World War I. Hitler's vision of the People's Community clearly reflects the influence of these radical nationalists.[6]

Winter Relief not only promoted nationalism, but it also exemplified and promoted Hitler's brand of socialism. In April 1934 he congratulated the leaders of the Winter Relief Drive, remarking that the program would "contribute to educating the *Volk* in socialist thinking."[7] Three years later he met again with the leaders of the Winter Relief Drive after the campaign finished. He thanked them for their efforts, since the Winter Relief was "a crucial instrument in the educational process of turning the German *Volk* into a true socialist community."[8] While initiating the fifth annual Winter Relief Drive in October 1937 he equated the campaign with true socialism. By showing care and concern for fellow Germans, those who were well-off would exercise the spirit of sacrifice. This would remind those who were less fortunate that they were part of a larger entity. The poorer segments of society would feel "sheltered in their *Volksgemeinschaft* [People's Community]."[9] The Winter Relief Drive would thus forge unity among Germans, insuring the strength and vitality of the German People's Community. When he opened the 1938 Winter Relief Drive he bragged that the success of the program would "prove beyond all doubt that the word '*Volksgemeinschaft*' is not just an empty delusion."[10]

Winter Relief built solidarity not only by appealing to the more impoverished members of the People's Community, but also by inculcating feelings of sympathy in those contributing. When he initiated the Winter Relief Drives in 1939 and 1940, he explained that if distributing money to the poor had been the only goal, then the government could have used taxation to fund poor relief. In many respects, this would have been simpler. However, using private, allegedly voluntary, contributions created a sense of community in a way that taxation would not.[11]

Another way Hitler and the Nazis fostered consciousness of solidarity with the poor was by encouraging all Germans to eat an inexpensive stew on the first Sunday of every month. The ten pfennig they saved by eating more frugally could then be contributed to the Winter Relief Drive. Hitler noted that some Germans did not appreciate Stew Sundays and preferred just to give the ten pfennig per meal, while eating as they pleased. However, this was simply not acceptable, Hitler intoned, since the reason for Stew Sunday was not merely to save funds to contribute to the poor, but also to experience privation for the sake of others. "We hold that, by such visible demonstrations, we are continually stirring the conscience of our *Volk* and making each of you once more aware that you should perceive yourself as a *Volksgenosse* [racial comrade], and that you should make sacrifices!" Thus, the point of Stew Sunday was not just to raise funds for the poor, but to demonstrate unity based on sacrificial giving. In addition, it made everyone experience deprivation, so they could better commiserate with the poor.[12] Hitler observed Stew Sunday himself, and he made sure the German public knew that he did.

This drive to raise funds for the poor epitomized the Nazi attempt to build a People's Community. In order for the German people to win the racial struggle, they needed to band together in social solidarity. Hitler's brand of National Socialism was supposed to build a powerful community that could withstand the onslaught of other competing races and also—as we shall see later—go on the offensive against surrounding nations to win more territory. As Peter Fritzsche has argued, the Nazi stress on building a united People's Community won Nazism many adherents in the run-up to its seizure of power. Many Germans earnestly desired national unity and a recovery of national power. However, as Fritzsche points out, few understood the expansionist implications of national unity in Hitler's social Darwinist worldview.[13]

Hitler often linked his building of the People's Community with the need to make Germany powerful against its enemies. Sometimes he even explicitly linked the People's Community with unity in the racial struggle. Just a few days after coming to power in Germany Hitler addressed the German nation, setting forth his primary goals. One key aim was preservation of the German *Volk* in its struggle for existence. He stated, "Because we perceive our highest goal to be the preservation of our *Volk*, enabling it to undertake its own struggle for existence, we must eliminate the causes of our own disintegration and thus bring about the reconciliation of the German classes." He saw the Marxist class struggle as a source of weakness for Germany, because it divided the nation, making it weaker in the struggle for existence and thus threatening its future existence and expansion. Racial unity in the People's Community would help Germany win the struggle for existence against the surrounding peoples.[14]

Hitler had already articulated essentially the same point in a speech before he came to power. In July 1932 he stated,

> A faithful community of people [*Menschen*] has arisen which will gradually overcome the prejudices of class madness and the arrogance of rank. A faithful community of people which is resolved to take up the fight [*Kampf*] for the preservation of our race, not because it is made up of Bavarians or Prussians or men from Württemberg or Saxony; not because they are Catholics or Protestants, workers or civil servants, bourgeois or salaried workers, etc., but because all of them are Germans.[15]

Overcoming class and religious divisions in German society, then, was not a means to promote human equality, a principle that Hitler decisively rejected. On the contrary, it was a method to increase national power. Hitler considered it a prerequisite to succeed in the universal racial struggle.

As we shall see in greater detail later, part of winning this racial struggle for existence meant expanding Germany's borders to accommodate a growing population. Hitler recognized that his expansionist program necessitated a Germany that was united and internally strong. In his 1939 May Day speech he explained that a strong People's Community was a prerequisite for gaining territory:

> The foundations for the life of a people are not to be found in doctrines and theories, but in its *Lebensraum* [living space], in what the earth affords it for sustenance. Hence, Lebensraum cannot be regarded separately from the *Lebenshöhe* (peak of life) of a *Volk*. And this Lebensraum is not enough by itself—and this also is a truly revolutionary realization—it must be complemented by a Volk's diligence, its energy, and its ability to manage to get the most out of its Lebensraum. And a still greater insight: this necessitates a *Volksgemeinschaft* [People's Community].[16]

So, gaining and utilizing new living space was dependent on forging and maintaining comradeship among the German *Volk*. Seven months earlier Hitler had told construction workers building fortifications in the Rhineland region that successful foreign policy required the inner unity of the People's Community.[17] For Hitler, national unity was a means to his expansionist ends.

Hitler's Socialism and Economics

Nationalism and socialism were inextricably linked in Hitler's mind and he often claimed the two were identical.[18] The very name of his party, which Hitler had changed in 1920 from the German Workers' Party to the National Socialist German Workers' Party, reflected this identification of nationalism and socialism.[19] For Hitler socialism did not mean economic equality or public ownership of the means of production, but rather it meant an economic system characterized by each working for the sake of the whole nation. Each should sacrifice his or her time, energy, and material goods to promote the common welfare of the German nation and people (defined racially, of course).

One reason that socialism was linked in Hitler's mind with nationalism and racism was because he believed that only the Aryan race had the innate moral qualities to implement socialism. We already examined his view of the Aryans as possessing the highest moral traits (see chapter 4). In his 1920 speech, "Why We are Anti-Semites," where he outlined his position on the Aryans as the sole producers of culture because of their innate tendency to hard work, he stated, "We were of the conviction that socialism in our sense can and will only be found with nations and races that are Aryan, and thus in the first place we hope in our own *Volk* and are convinced, that thereby socialism is indivisible from nationalism."[20] Two years later he reiterated this basic position, asserting that socialism "has indeed grown up exclusively in Aryan hearts."[21]

If the Winter Relief Drive was a prime example of Hitler's socialism, then it is apparent that socialism meant something quite different to Hitler than it did to many of his contemporaries. Most socialists in the early twentieth century would have dismissed voluntary contributions to help the poor as far too little. Rather, most socialists preferred government ownership of the means of production to alleviate the miserable conditions of the working class. Indeed, even though Hitler often criticized capitalism, big business, interest, and unearned income as tools that immoral Jews used to fleece hardworking Germans, he did not advocate an abolition of private property, as Marxists did.

In January 1932 Hitler explained to the Düsseldorf Industry Club why he believed in private property. First, he instructed them about his worldview, which rested on three foundations: (1) the hereditary value of the *Volk*; (2) the value of personality (*Persönlichkeitswert*); and (3) the universality and necessity of struggle. When Hitler spoke about the "value of personality"—which he did quite often[22]—he meant that individuals within the *Volk* were biologically unequal, possessing different hereditary aptitudes and talents. The achievements of any given nation or people were the product of the most intelligent and creative individuals in that society. Only those of superior abilities crafted civilization and higher culture, so they should be allowed to rise to higher positions in society than the less able masses.

Hitler drew both political and economic conclusions from this belief in human inequality. Politically, democracy presupposed the equality of individuals, so it must be superseded by the leadership principle (*Führerprinzip*). The problem with democracy, he asserted, is that "when the capable minds of a nation—who are always in the minority—are given a value equal with all the others, this must result in subjugating the genius to the majority, in subjecting the ability and the value of the individual to the majority, a process which is mistakenly called the rule of the people." Rather than submit to this "rule of stupidity," as he called it, a people should be ruled by "its most capable individuals who are born for the task."[23]

Economically, the difference in individual performance implied private property, Hitler told the industrialists of Düsseldorf. Private property, he stated, "must draw its ethical justification from the insight that it is a necessity dictated by nature." What was this law of nature that made private property necessary? Essentially it was the principle of personality that he had already mentioned. He explained that "private property is only morally and ethically justifiable if I assume that men's achievements are different." It would be illogical, he continued, to assign the fruits of one individual's achievements to those with lesser abilities.[24] Those individuals with the inborn talents and abilities to succeed economically should have control of the assets they gain in open competition.

Thus, Hitler based economic inequality on biological inequality. He explained this quite clearly in his closing speech to the Nazi Party Rally in Nuremberg in September 1933: "The conception of private property is thus inseparably connected with the conviction that the capacities of men are different alike in character and in value and thus, further, that men themselves are different in character and value."[25] These utterances were not just propaganda to gain capitalist support for the Nazi Party. Hitler confided to a close associate that egalitarianism was flawed. He stated, "A process of selection must be introduced in some way if one wishes to arrive at a natural, healthy, and satisfactory solution to the problem—a selection process for those who have or should have a claim and right to ownership and proprietorship of any business."[26] The selection process, of course, was open competition, decided not by inherited privileges—whether aristocratic or capitalistic—but by talent and ability inherited biologically.

Hitler's idea that biological ability should prevail in free economic competition was a form of social Darwinism that had been prevalent in Germany and elsewhere for decades. Darwin himself had embraced this idea in *The Descent of Man*, where he criticized the custom of allowing the firstborn son to inherit the estate, because this conferred economic advantages that were not based on biological qualities. Darwin insisted that economic competition was the best way to promote evolutionary progress.[27] In a private letter Darwin lamented that unions and cooperative societies reduced competition. This curtailing of economic competition "seems to me," he remarked, "a great evil for the future progress of mankind."[28] Darwin construed economic competition as a beneficent process contributing to higher evolution.

Darwin's concern that economic competition was the best path to biological progress was shared by many German social Darwinists in the late nineteenth century. Prominent biologists, such as Haeckel, Wilhelm Preyer, and Alexander Ecker, integrated laissez-faire economic competition into their writings about Darwinism.[29] Haeckel, Preyer, Oskar Schmidt, and other Darwinian biologists lambasted socialism as unscientific, since it allegedly contradicted the Darwinian assumption of biological inequality of organisms, including humans.[30] The famous scientific materialist and tireless popularizer of Darwinism, Ludwig Büchner, espoused a form of socialism based on the Darwinian struggle for existence. Büchner's socialism was founded on the idea that everyone should start off life without significant economic advantages over their fellow men and women. Thus he opposed the inheritance of capital and property. However, he also strongly criticized Marxism for its allegedly un-Darwinian idea of human equality. Büchner believed strongly that human inequality was a necessary condition for evolution. By providing everyone equal economic opportunity, biological talent and ability would prevail. This would allow those with superior biological qualities to succeed economically, thus contributing to evolutionary progress.[31] Many biologists and social thinkers promoted similar views before and during Hitler's time.

In the early years of the Nazi Party, Hitler had seemed more socialist—or at least more anticapitalist—than he did later on. The Twenty-Five Point Program of the Nazi Party had boldly proclaimed that Nazism stood for abolition of unearned income, "breaking the bondage of interest," nationalization of trusts, profit-sharing, replacing large department stores with small businesses, and land reform. However, Hitler never renounced private property per se. In his 1924 trial for treason, he criticized Marxism for refusing to acknowledge the "value of personality," which was a fundamental principle of human nature and the driving force behind history. He also claimed that the principle of personality was the basis for private property, so Marxism erred by wanting to abolish private property.[32]

Unlike the racial policies outlined in the Twenty-Five Points, which were later implemented during Nazi rule, the Nazi economic program was flexible. Hitler was only concerned about the economy to the extent that it facilitated or hindered the preservation and reproduction of the German people.[33] The second chapter of his *Second Book* was entitled, "The Struggle, Not the Economy, Secures Life." Therein he argued that the economy is only important to the extent that it helps the *Volk* in its struggle for existence.[34] His socialism was always a tool for a higher purpose. He did not always make that clear in the early days of the Nazi Party, since he hoped to use the socialist planks in the Nazi Party program to attract workers and forge a mass movement. However, after coming to power he laid his cards on the table at a May 1934 meeting of the German Labor Front. "Socialism cannot exist for Socialism's sake," he told them. "A revolutionary upheaval can be justified only if in fact in the final result it serves to advance a people's self-preservation and the preservation of its life. That is the sole justification for Socialism."[35]

Hitler's brand of socialism was a means to an end, and the end was the perpetuation of the German *Volk*. His socialism also aimed at promoting evolutionary progress, as was made clear in the booklet Hitler commissioned, *Why Are We Fighting?* (1944). This booklet explicitly linked socialism with evolution, stating, "Socialism means for us not the solution of the labor question, but rather the ordering of all German racial comrades into a genuine living community; *it means the preservation and further evolution of the Volk on the basis of the species-specific laws of evolution.*"[36] For Hitler, socialism served the purpose of biologically advancing the German *Volk*.

Thus, for Hitler economics was always subservient to racial policy. It was integrally related to foreign policy, since he believed that territorial expansion was the only viable long-term solution to provide for an expanding German population. At the Nuremberg Party Rally in 1929, Hitler informed his enthusiastic followers that the economy was not his primary concern. Improving economic conditions would not solve society's problems. Rather, his focus was on restoring the hereditary health of the German people.[37] Hitler was never all that concerned about the German standard of living. For him the economy was just a means to provide sustenance for an expanding population, as well as a means to build military power to gain new agricultural land. Because Hitler wanted the economy to be subservient to racial and military interests, however, he did not hesitate to intervene in the economy to direct it to those ends.[38]

Darwinian Politics and the Leadership Principle

Although Hitler only occasionally forthrightly linked his ideas about biological inequality and the struggle for existence to private property, as he did in his speech to the Düsseldorf industrialists, he often applied these Darwinian principles to the political process. He explained his inegalitarian political philosophy at length in *Mein Kampf*, where he asserted,

> *A philosophy of life which endeavors to reject the democratic mass idea and give this earth to the best people—that is, the highest humanity—must logically obey the same aristocratic principle within this people and make sure that the leadership and the highest influence in this people fall to the best minds. Thus, it builds, not upon the idea of the majority, but upon the idea of personality.*[39]

Giving the earth to the "highest humanity" would foster the upward evolution of humanity, so even his leadership principle flowed from evolutionary ethics. Haeckel had already appealed to the "aristocratic principle" to oppose the democratic and egalitarian ideals of socialism. In 1895 Haeckel wrote, "Darwin's theory of selection is closely linked with the biological laws of the division of labor; it is not democratic, but rather an aristocratic principle."[40] Whether Hitler learned about the "aristocratic principle" directly from Haeckel or indirectly through Haeckel's disciples, he certainly made liberal use of it in his writings and speeches.

Further, Hitler regularly used explicitly Darwinian terminology to justify his leadership principle (*Führerprinzip*), which specified that the masses should submit to those leaders who had proven their ability to guide the nation. These able leaders would emerge through a competitive process, he explained: "The selection of these [best] minds, as said before, is primarily accomplished by the hard struggle for existence."[41] While lambasting the parliamentary system and extolling the virtues of leadership by able and responsible individuals, Hitler complained that "the parliamentary principle of majority rule sins against the basic aristocratic principle of Nature."[42] Natural law was thus his source for morality, and since he believed that inequality is engrained in nature, he strived to bring human society into conformity to natural inequalities. Hitler was not thereby endorsing an aristocracy of birth, but an aristocracy of talent. Hitler, then, justified the leadership principle by appealing to the laws of nature, just as he grounded all his moral tenets on principles derived from nature, especially evolutionary principles.[43]

In a private speech to Nazi officials in 1937, Hitler expatiated on the glories of the leadership principle. Identifying the right leaders should take place "through a natural selection." This selection process was easy in the "time of struggle" (*Kampfzeit*), that is, in the early years of Nazism, because struggle always produces the best selection, Hitler averred. He was obviously patterning the political process after biological evolution. In order to produce the best leadership possible in the future, this "natural selection process" must begin in youth.[44] In a proclamation of September 1938, Hitler declared that the Nazi leadership had been built up through a "ruthless selection process" that had functioned especially well during the "time of struggle" (*Kampfzeit*).[45] This message was especially welcome to the "old fighters" (*Alte Kämpfer*) in the Nazi Party, whom Hitler was hereby designating as superior. In 1938 he told the German parliament that the Nazi Party had organized the nation in such a way "that the supremely natural principle of selection would appear to indicate that the continued existence of a secure political leadership is guaranteed."[46] This "natural principle of selection" was, of course, a social Darwinist competitive ethos. Hitler wanted to structure society to allow the born leaders to succeed in the struggle for existence and then take command.

Hitler's meritocratic socialism was a way to make Germany stronger and more efficient by producing talented political leaders and capable business leaders who would work, not for their own interests, but for the welfare of the entire German people. In the first year of his rule Hitler defined socialism as a system that makes sure that "on each man should be placed that share in the maintenance of the people as a whole which corresponds with his inborn talent and his value."[47] A few months earlier he had defined socialism similarly, stating, "Socialism is nothing else than the natural ordering of a people according to its inborn capacities."[48] Finding one's place in the People's Community through competition would benefit the whole nation, Hitler thought.

It is unclear if Hitler's penchant for political and economic competition or his own lack of organizing skill—or both—produced the polycracy that seemed to characterize Nazi leadership. Heinz Linge, Hitler's adjutant, claimed that Hitler consciously gave a free hand to other Nazi leaders, causing them to fight among themselves. Linge even called this process "Ämterdarwinismus," meaning a Darwinian competition for offices or between offices. Linge insisted, however, that Hitler ultimately held the reins.[49] It seems most probable that Hitler's underlings, though often fighting among themselves, were always "working towards the Führer," as Kershaw has emphasized. Also, it is clear from many sources, especially Goebbels' diaries, that Hitler intervened frequently in the decision-making process in the Third Reich, not only in matters of the highest urgency, but often in affairs that seem trivial. Thus, whether the polycracy was a conscious ploy or not, he used the competition among his colleagues to serve his own purposes.

Primacy of Community over the Individual

The essence of the People's Community was encapsulated in the famous Nazi motto, "common welfare before individual interest." Suppression of individual interests for the sake of the nation or race epitomized Nazi morality. Laying down one's own desires and pleasures for the sake of others sounds very altruistic, and Nazi ideals did have a façade of virtue. Several years before coming to power, Hitler, sounding very moral, criticized individualism, confiding to Otto Wagener, "In the socialism of the future, on the other hand, what counts is the whole, the community of the *Volk*. The individual and his life play only a subsidiary role. He can be sacrificed—he is prepared to sacrifice himself should the whole demand it, should the commonweal call for it."[50]

The idea that the individual counts for nothing was commonplace among Darwinists in the late nineteenth and early twentieth centuries. One of the most famous Darwinian biologists in Germany, August Weismann, stated in 1881 that "*only the interest of the species* comes into consideration, not *that of the individual*."[51] The following year Büchner similarly commented, "The individual is nothing in relation to the course [of time], the species is everything."[52] By the early twentieth century many eugenicists were arguing this same point rather forcefully. For example, Wilhelm Schallmayer contended that evolution shows that individual interests are only significant inasmuch as they contribute to the welfare of the species: "This natural law, the complete subordination of the individual interests under those of the species, must also be valid for human evolution." Schallmayer criticized European culture for laying too great a stress on the value of the individual, which sometimes damages the interests of the species. Schallmayer and many other eugenicists thus devalued the lives of individuals, making their value dependent on their contribution to evolutionary progress.[53]

Hitler's belief that the collective (species and/or race) takes priority over the individual thus had a long pedigree among social Darwinists and eugenicists in Germany. However, one of the key differences between Nazi virtue and most other conceptions of altruism was that if an individual did not willingly sacrifice himself or herself for the sake of the community, the Nazis were ready and willing to exercise compulsion. In this respect, even the police forces were supposed to be an instrument to shape the People's Community. Hitler encouraged the German police in September 1937 to be the best friend of

Figure 5.2 Nazi Weekly Proverb quoting Hitler: "The individual must and will as always perish; only the Volk must remain."

the German *Volk*. However, they also had the task of "being the most relentless representative of this People's Community toward those asocial, criminal elements which sin against it."[54] Their ruthless crushing of dissent and brutality toward political dissidents and those of other races were interpreted by the Nazis as acts of service for their racial comrades.

Individual liberties were always subservient to the interests of the People's Community. Hitler expressed this view often, including this clear statement in a 1939 speech: "The liberty of the individual ends where it starts to harm the interests of the collective. In this case the liberty of the *Volk* takes precedence over the liberty of the individual."[55] In his second May Day speech Hitler put a positive spin on his ruthless crushing of trade unions, political parties, and other non-Nazi organizations the previous year. Eliminating these competing organizations contributed to the inner unity of the German people. "We have thereby redeemed the German people from an endless amount of inner strife and wrangling," he rationalized.[56]

Though Hitler clearly believed in human biological inequality that would produce political, economic, and social inequalities, he did embrace one kind of equality: equality of opportunity.[57] Like Büchner and some other moderate socialists before him, his socialism was a meritocratic system that allowed each individual to succeed or fail based on his or her own qualities and abilities. Rather than inherited capital or aristocratic privileges determining the course of one's life, the struggle for existence should decide one's fate. This is why Hitler was so insistent on providing universal education for all young Germans and free higher education to those of ability. In *Mein Kampf* Hitler wrote a section on the need for educational opportunity for all, including the poor. He opened the section by stressing the need for the state to practice "human selection." This selection based on ability would help sweep away class divisions, because success would depend entirely on ability, not on wealth or social status.[58]

Once the Nazis came to power, they continued offering free education to all those deemed intellectually capable and talented. However, they introduced a novel prerequisite to higher education: physical health. After requiring medical exams for all new university students (which eugenicists had already introduced at some universities before 1933), the Education Ministry ordered in 1935 that chronically ill students be excluded from the universities.[59] The Nazis also chose students they considered exemplary (with criteria changing over time) to attend an elite Adolf Hitler School in Sonthofen. In a secret speech to military leaders in June 1944, Hitler boasted that the Sonthofen school was a prime example of the process of selection that the Nazis promoted.[60]

As in every other arena, Hitler promoted education because he considered it advantageous in the struggle for existence against other nations and races. In *Mein Kampf* he articulated this position forthrightly:

> The state has the obligation to exercise extreme care and precision in picking from the total number of national comrades the human material visibly most gifted by Nature and to use it in the service of the Community.... Another factor for the greatness of the people is that it succeed in training the most capable minds for the field suited to them and placing them in the service of the national community. *If two peoples, equally well endowed, compete with one another, that one will achieve victory which has represented in its total intellectual leadership its best talents and that one will succumb whose leadership represents only a big common feeding crib for certain groups or classes, without regard to the innate abilities of the various members.*[61]

Even Hitler's educational program was designed to increase German power vis-à-vis its neighbors. It would render them more competitive in the inevitable national and racial contests.

Nazi Welfare and Biological Inferiority

As Richard Evans has pointed out, there was an inherent tension between Hitler's social Darwinist worldview and charitable efforts. Indeed, Hitler often disparaged Christian charity and humanitarianism, because it helped the weak to survive and propagate their bad heredity.[62] However, Hitler was not being entirely inconsistent by supporting some kinds of welfare, because he and his regime always tried to distinguish between those in society who were hereditarily weak and those who were weak because of the oppressive capitalist economic system (which he blamed on the Jews). While disparaging *"philanthropic flim-flam"* in *Mein Kampf,* he proposed a twofold path to improve social conditions in Germany: "The deepest sense of social responsibility for the creation of better foundations for our evolution, coupled with brutal determination in breaking down incurable tumors."[63] Nazi charity was aimed at helping those who were deemed valuable members of the People's Community, but who were disadvantaged by the capitalist system.

However, Nazi welfare programs were not intended to help those who were considered inferior biologically, those whom Hitler so contemptuously called "*incurable tumors*." This was clear from the start.[64] At the same time that Hitler announced the first Winter Relief Drive, the Nazis also initiated Beggars' Week, during which police swept through cities, removing vagrants, whom Nazis considered biologically inferior.[65] In 1938 the Gestapo put over ten thousand vagrants and beggars in concentration camps, since by that time everyone chronically unemployed was labeled work-shy and "asocial."[66] Not only vagrants and the unemployed, but also alcoholics, prostitutes, and "habitual criminals" were often labeled "asocial" by the Nazis.

Many biologists and eugenicists before and during the Nazi period thought that "asocial" characteristics were hereditary traits. The geneticist Siegfried Koller coauthored with Heinrich Wilhelm Kranz, director of the Institute for Genetics and Racial Hygiene at the University of Giessen, a major study of "asocials" in 1939–1941. They wrote in their book, "If we want to move toward a biological solution of the antisocial problem, it is absolutely imperative to deem antisocials from antisocial families as the biologically most unhealthy and most dangerous for the people." They advocated compulsory sterilization, forced labor, marriage prohibitions, and annulment of existing marriages for those deemed "asocial." Koller's work was well-received by Nazi authorities, who appointed him to the Biostatistical Institute of the University of Berlin in March 1941. They asked him to oversee planning to solve the problem of "asocials."[67]

The view that "asocials" were biologically inferior people unworthy of receiving state welfare was reflected in the draft legislation prepared in 1944 by the Nazi regime. The "Law about the Treatment of *Gemeinschaftsfremder*" never went into effect, but it mirrored Nazi attitudes toward the "asocial," who by 1939 had been dubbed "*gemeinschaftsfremd*." This term means literally "foreign to the community," but it implies that these people are unable to fit into society, as another Nazi term for them, "*gemeinschaftsunfähig*," more clearly indicates. The Nazi justification for the draft law also made clear that these "*Gemeinschaftsfremder*" were biologically incapable of integrating into the People's Community. Therefore, they should not be accorded state assistance, but rather should be subject to police measures, which presumably meant internment.[68] The Interior Minister

Frick issued a directive to German Health Offices in July 1940 telling them that progeny from "asocial" people was undesirable, though he did not specify what practical measures should be taken to prevent their reproduction.[69]

Nazi charity was limited to those considered healthy members of the People's Community. This was apparent in the practices of the National Socialist People's Welfare Organization, too. From the start this party organ distributed welfare based on racial and biological criteria. It also aimed to replace the Catholic and Protestant charitable organizations, which Nazis considered detrimental, because they helped even the biologically "unfit."

Another group Hitler excluded from the People's Community was those who did not work. While many private charitable organizations refused to support those who were able but unwilling to work, Hitler excluded even those who were unable to work. Speaking to a party meeting in March 1927, Hitler stated, "I only act in the interest of a *Volk*, if with the highest zeal I endeavor to preserve the life of an entire *Volk*, insofar as it is profitably employed and valuable. Who then is valuable? Valuable is every person who at the expense of and on the basis of his ability works and produces for the *Volk*."[70] Those who will not work—as well as those who cannot—were thus not considered valuable members of the community. They were excluded from the rights and privileges of the community, including even the right to life. In April 1923 Hitler stated, "In the People's Community the only one who has a right to live is the one who is prepared to work for the People's Community."[71] As we shall see, eliminating the right to life for those unable to work had dire consequences: In 1939 Hitler ordered the killing of the hereditarily ill, of those who could not labor for the national community.

Like other social Darwinists before him, Hitler had to deal with the inherent tension between the individual struggle for existence taking place within society and the struggle for existence between peoples and races. Most social Darwinists thought that the struggle was occurring on both planes simultaneously, but some stressed one level more than the other.[72] In Hitler's case, he clearly considered the racial struggle more significant than the internal struggle. With the prominent exceptions of those Germans identified as non-Aryans or those having hereditary illnesses, the internal struggle was supposed to be a peaceful economic and political competition for position and influence that would result in a harmonious, efficient society. The racial struggle, on the other hand, was a battle to the death for territory and nourishment. Those nations who set aside class differences to unite against their racial enemies would succeed best and would ultimately destroy those without such cohesion. Hitler's socialism, his building of the People's Community, and Nazi welfare policies were all intended to unify and strengthen the German body politic (*Volkskörper*) biologically and militarily, so it could triumph against other allegedly inferior races in the struggle for existence.

CHAPTER SIX

Sexual Morality and Population Expansion

The Centrality of Reproduction

On May 6, 1933, four days before the Nazi's famous episode of book-burning at the German universities, when they consigned volumes by Marxists, Jews, and modernists to the flames, the Nazis staged a preview in Berlin. They ransacked the Institute for Sexual Science led by the famous gay activist and sexologist Magnus Hirschfeld, who had left Germany before the Nazis came to power and wisely stayed abroad afterward. The Nazis torched over 10,000 volumes from the institute's library, along with a bust of Hirschfeld. Boasting about these exploits in their Berlin newspaper, the Nazis proclaimed their resolve to clean up the smut purveyed by Hirschfeld and his ilk.[1] Many conservatives applauded Hitler's campaign to cleanse German culture of erotic filth, and many scholars today see Nazi sexual morality as antimodernist. In some ways it was. However, few people then (or now) really understood Hitler's sexual morality, which did not fit comfortably within the moral categories of most of his fellow Germans.[2]

Hitler's greed for total control of German society affected the most intimate aspects of human life. As a biological determinist endeavoring to improve the hereditary quality of the German people, controlling reproduction was a central concern of his. Not all Germans understood how determined Hitler was to control their sexual lives, but in a secret speech in November 1937 Hitler informed Nazi Party leaders: "Today *we* are laying claim to the leadership of the *Volk*, i.e. we alone are authorized to lead the *Volk* as such—that means every man and every woman. The lifelong relationships between the sexes is something *we* will regulate. *We* shall form the child!"[3] A few months earlier Himmler had made a similar point in a speech, when he stated that "all things which take place in the sexual sphere are not the private affair of the individual, but signify the life and death of the nation."[4] Obviously, the Nazi regime never attained such absolute control over the sexual lives of their citizens as perhaps Hitler or Himmler would have wished. However, they did make significant inroads.

Hitler's sexual morality comported in many ways with conservative values, since he opposed birth control, abortion, and homosexuality. He pilloried the erotic culture of urban centers and hoped to throttle prostitution. Hitler posed as a moral crusader for conservative values by pledging to eliminate sexually explicit content from German culture. Nazi propaganda consistently portrayed Hitler and the Nazi Party as upholders of family values and clean morals. However, as Jill Stephenson has pointed out, Nazis' morality "was nothing short of a revolution, which would drastically alter the nature, if not threaten the existence, of the family unit which the Nazis had originally pledged themselves to protect and promote."[5] Protecting the family was only a priority with Hitler if it helped him achieve a different mission.

Ultimately Hitler's sexual morality differed in important respects from the values dear to many of his conservative supporters.[6] His goal of biological improvement of the German people and his willingness to countenance any means to reach this end placed him at odds with many social conservatives. Thus, while the Nazi regime promoted early marriage, it also relaxed divorce laws, encouraged extramarital sexual affairs, and tried to eliminate the stigma of illegitimacy. While banning contraception and incarcerating homosexuals, the Nazis also introduced sweeping new marriage restrictions and compulsory sterilization. Though prohibiting most abortions, they compelled some women to have abortions. During World War II, Hitler and Himmler even discussed allowing polygamy after the war. These seemingly disparate and even contradictory policies were not as disconnected as it seems on first glance. Hitler and his colleagues intended all these policies to serve a common purpose: to improve the German people biologically. Hitler told Goebbels in December 1940 that sexuality cannot be regulated according to Christian hypocritical morality; rather, "We must view this question [of sexual morality] entirely from the standpoint of the usefulness for the *Volk*. That is our morality."[7]

In some cases, the Nazis could not decide what was most advantageous biologically. Nazi policy relating to prostitution, for example, altered during the Nazi period, as officials changed their minds about what was most conducive to promote the health and fitness of the German nation. At first, Nazis tried to squelch prostitution to stymie the spread of syphilis. They branded prostitutes as "asocial" and incarcerated thousands of them. However, by the mid to late 1930s Nazi officials, including Himmler, came to believe that state-run brothels were needed, especially for soldiers and SS men away from their families. Once war broke out, the Nazi regime began establishing brothels for the military. Later the SS even established brothels for concentration camp inmates.[8] For Hitler prostitution was only evil if it infected Germans with sexually transmitted diseases, because he believed—as did many biologists and physicians at the time—that these diseases caused biological degeneration in gametes and thus in the offspring. They encouraged prostitution if they thought it would advance their objective of biological health and vitality for the German nation.

Since Hitler's chief goal was to further human evolution, he bent all moral precepts, including sexual morality, to serve this end. While his sexual morality converged with the Christian sexual mores dominant in his society at some points, they sharply diverged in many places. At times Hitler overtly criticized Christian sexual morality, and he certainly never justified his own sexual morality by appealing to scripture or religious tradition. The guiding principle behind his sexual morality was evolutionary ethics. Ultimately, Hitler defined as morally good any sexual activity that contributed to evolutionary progress. Sexual sins were relationships that produced "inferior" offspring.

Sexual morality was not peripheral to his worldview, either, since Hitler knew that the ultimate winners of the Darwinian struggle for existence were those who could reproduce more prolifically than their competitors. In the opening passage of his *Second Book* he claimed that all politics and history are driven by the human struggle for existence, which is inescapable, because the two mightiest drives motivating humans are hunger and love. "In truth," he stated, "both these drives are the rulers of life." Hitler considered survival and reproduction the basic motivation for all human behavior, both for individuals and for races.[9] He insisted that all his own policies were motivated by concern for the survival and reproduction of the German *Volk*, which was justified because they were the highest and best people on the earth.[10] In a 1927 speech, he explained the primacy of reproduction in politics:

Politics is the striving and struggle of a *Volk* for its daily bread and its existence in the world, just as the individual devotes its entire life to the struggle for existence, for its daily bread. And then comes a second matter: caring for future survival, caring for the child. It is the struggle for the moment and the struggle for posterity. And all thinking and all planning serve in the deepest sense this struggle for the preservation of life.[11]

According to this perspective, reproduction is of the utmost importance, affecting all other considerations.

In *Mein Kampf* Hitler frequently discussed reproductive issues. He wanted attention to "be directed on increasing the racially most valuable nucleus of the people and its fertility, in order ultimately to let the entire nationality partake of the blessing of a highly bred racial stock." He then called for government measures to ensure this. Ultimately in his ideal state

> *the folkish philosophy of life must succeed in bringing about that nobler age in which men no longer are concerned with breeding dogs, horses, and cats, but in elevating man himself, an age in which the one knowingly and silently renounces, the other joyfully sacrifices and gives.*[12]

Hitler manifested a utopian impulse to improve humanity biologically by controlling reproduction and breeding better humans. He used the morally loaded language of renunciation and sacrifice to describe the sexual activities of those renouncing child-bearing and those sacrificially bearing children to produce this higher breed of people.

In this passage Hitler implied that planned reproduction would be voluntary in his utopian society. However, he also approved of government intervention, proclaiming that the state "must declare the child to be the most precious treasure of the people." He explained that those who are unhealthy should forgo reproduction, while "it must be considered reprehensible: to withhold healthy children from the nation." He continued, "Here, the state must act as the guardian of a millennial future in the face of which the wishes and the selfishness of the individual must appear as nothing and submit."[13] Hitler thus tried to stake claim to the moral high ground, accusing anyone who would object to his repressive sexual policies as selfish and unconcerned for the welfare of the community. He also implied that such efforts at controlling reproduction would result in a "millennial future." Hitler's thousand-year Reich depended on breeding a higher humanity by enforcing a new sexual morality in German society and inculcating the youth with these new ideals.

Increasing the Birthrate

One major concern of Hitler's sexual morality was increasing the birth-rate in Germany. This was motivated to some extent by long-range military considerations: more babies meant more future soldiers. He justified and encouraged population expansion as a source of national strength. In February 1934 he expounded to some close associates that Germany needed to increase its population in order to be in a stronger position to carry out its foreign policy.[14] In January 1942 he told some close associates that he often contemplated the reasons for the fall of ancient civilizations. He believed that the primary reason for their decline was because their upper classes ceased bearing as many children. "Our salvation," he asserted, "will be the child!"[15] He also criticized the French for limiting their population, thus bringing about stagnation and decline. Hitler equated a large population with power, and he wanted Germany to become even more powerful.

However, building a larger military in the future was not the primary reason Hitler favored population expansion. Most of the time when he explained why he considered population expansion essential—as he did often—he emphasized the need to maintain or improve the biological quality of the German people. In *Mein Kampf* Hitler discussed at length ways to deal with the expanding population of Germany, which was increasing, according to his figures, by 900,000 people per year (this is the same figure that Heinrich Class gave in his bestselling book, *Wenn ich der Kaiser wär*). One possibility he rejected was to practice birth control to limit the population. This solution, he claimed, would ultimately fail, because it violates the laws of nature. Nature, he remonstrated, follows "a method as wise as it is ruthless," since it restricts population growth

> by exposing them to hard trials and deprivations with the result that all those who are less strong and less healthy are forced back into the womb of the eternal unknown. . . . By thus brutally proceeding against the individual and immediately calling him back to herself as soon as he shows himself unequal to the storm of life, she keeps the race and species strong, in fact, raises them to the highest accomplishments.[16]

By producing many individuals and then selecting the best ones through competitive struggle, nature ensures that a species maintains and even improves its biological quality.

When humans artificially restrict births, they no longer follow this wisdom of nature, Hitler thought, because they preserve everyone who is born, regardless of biological quality. Hitler hammered this point home:

> For as soon as procreation as such is limited and the number of births diminished, the natural struggle for existence which leaves only the strongest and healthiest alive is obviously replaced by the obvious desire to "save" even the weakest and most sickly at any price, and this plants the seed of a future generation which must inevitably grow more and more deplorable the longer this mockery of Nature and her will continues.[17]

Limitation of births, then, would lead to biological degeneration.[18] Thus Hitler based his opposition to birth control on evolutionary ethics. In his view the struggle for existence was a positive force in history, and birth control would diminish the beneficial effects of that struggle.

Hitler's belief in the necessity of population expansion was constant throughout his career. In a private speech to military officers in February 1942 he explained essentially the same point he had made over fifteen years earlier in *Mein Kampf*. He told them that nature dictated that populations expand and struggle for resources. The competitive process of natural selection would leave the best to inherit the earth. He drew the conclusion that birth control would undermine the beneficent effects of natural selection.[19]

Hitler made this point about the blessings of population expansion for improving biological quality yet another way in some of his speeches. He pointed out in a 1928 speech that if births were restricted, some great leader or inventor might never be born.[20] The following year he told the Nuremberg Party Congress that it was dangerous to set aside the process of natural selection by restricting births. His rationale was that the "first born are not the talented ones or the strongest people."[21] Hitler, of course, was not the first born in his family, so of course he did not think the firstborn were the greatest. In his *Second Book* he insisted that Germany's cultural achievements of the past would have been impossible if Germans had restricted their births. He stated, "If one were to strike out from our German cultural life, from our science—yes, from our entire existence—everything accomplished by men who were not firstborn, Germany would hardly even be at the level of a Balkan state. The German people would no longer possess any claim to being valued as a cultured people."[22]

Hitler's desire to foster reproduction colored his view of women's roles. Whenever Hitler addressed the Nazi Women's Organization or other female audiences, he taught them that their main purpose in life was reproduction. Hitler tried to convince German women that this was a noble duty incumbent on them to help Germany emerge triumphant in the universal struggle for existence. "Every child that she brings into the world," Hitler proclaimed in a 1934 speech to the Nazi Women's Organization, "is a battle which she wages for the existence or non-existence of her *Volk*." Thus, even during peacetime, Hitler believed that Germany's whole existence was at stake, and women needed to help fight the unceasing struggle by bearing children. Later in the same speech he told the women that "the program of our National Socialist Women's movement actually contains only a single point, and this point is: *the child*, this tiny being who must come into existence and flourish, who constitutes the sole purpose of the entire struggle for life."[23] Women played a crucial role in winning the struggle for existence, Hitler thought, since the people who reproduced most prolifically would ultimately win.

Hitler's concern about the effects of a declining birthrate on the biological vitality and evolutionary progress of the German people was a common theme in eugenics literature in the early twentieth century. In a book written shortly before World War I, the famous professor of hygiene and avid eugenics advocate, Max von Gruber, warned about biological degeneration that would occur if German birthrates continued to decline.[24] He voiced the same concern in a 1918 article in *Germany's Awakening Renewal* that Hitler may well have read.[25] Many other eugenicists, including Ploetz, agreed with Gruber that limitation of births would result in biological degeneration.[26]

Though Hitler and other Nazi leaders continually depicted their movement as supporters of the traditional family, they only supported the traditional monogamous family to the extent that it fostered their goal of population expansion and improvement of the species. Hitler never manifested concern about the extramarital affairs of his colleagues (unless it would damage the popularity of his regime).[27] However, generally he did favor monogamy as the best family structure for promoting reproduction. In an extensive passage in *Mein Kampf* on combating syphilis, he claimed that syphilis was dangerous to the health of the nation, since it might lead to infertility or biologically degenerate offspring. If the problem of syphilis were not solved, Hitler claimed, "the civilized peoples [would] degenerate and gradually perish." Hitler proposed early marriage as the primary antidote for syphilis.[28]

Figure 6.1 "The German Mother: Every child that she brings into the world is a battle that she wins for the existence or non-existence of her Volk. Adolf Hitler." (from Nazi periodical)

However, while promoting marriage and combating sexual profligacy, Hitler divulged the ultimate goals his proposals were supposed to serve. He stated that "marriage cannot be an end in itself, but must serve the one higher goal, the increase and preservation of the species and the race. This alone is its meaning and its task."[29] Thus, for Hitler marriage was not sacred, but was only a means to an end. What form marriage or sexuality should take were subsidiary to promoting evolutionary progress. If monogamy served the interests of the species and race best—as Hitler thought it did for the most part at the moment—then it should be promoted. If it ceased to advance the interests of the species, then it could be altered.

Hitler's support for monogamy mirrored the views of many leading eugenicists, including Ploetz and Gruber. Like them, his support for monogamy was based entirely on biological considerations. Interestingly, however, a few eugenicists in the early twentieth century dissented, proposing that polygamy would better advance human evolution. The philosopher Christian von Ehrenfels and the chemist and anti-Semitic publicist Willibald Hentschel were the most prominent advocates of replacing monogamy with polygamy. Other eugenicists, such as August Forel and some feminist eugenicists, pressed for freer sexual relations to replace strict monogamy. This debate among eugenicists over marriage reform was reflected in discussions among Nazi leaders about marriage and sexual relations, as we shall see.[30]

In the passage of *Mein Kampf* where Hitler discussed syphilis and the need for early marriages for men, he also criticized the proliferation of sexual indecency in Weimar culture, which promoted early sexual experiences and thus contributed to the spread of syphilis. He called for a purge of cultural life and wanted to "clear away the filth of the moral plague of big-city 'civilization,' " even if many Germans would oppose this. He hoped to eliminate eroticism from all forms of cultural life:

> Theater, art, literature, cinema, press, posters, and window displays must be cleansed of all manifestations of our rotting world and placed in the service of a moral, political, and cultural idea. Public life must be freed from the stifling perfume of our modern eroticism, just as it must be freed from all unmanly, prudish hypocrisy. In all these things the goal and the road must be determined by concern for the preservation of the health of our people in body and soul. The right of personal freedom recedes before the duty to preserve the race.[31]

While wanting to purge German culture of sexually explicit material, however, Hitler distanced himself from "prudish hypocrisy," thus distancing himself from moral conservatives.

Once Hitler came to power, he implemented policies to encourage early marriage and reproduction. One of the more popular policies was interest-free marriage loans of one thousand marks introduced in

June 1933. This was intended to help couples financially, so they could marry at a younger age. The marriage loan program also provided financial incentives for couples having children, since one-fourth of the loan was forgiven for each child born. By March 1937 about 700,000 German couples had taken advantage of the marriage loans. However, these couples only had one child on average, so the loans did not increase German birthrates as much as the Nazis had hoped it would.[32] The Nazi regime also introduced cash payments for children to encourage Germans to have larger families. In 1935 they offered families with over four children under age sixteen a onetime stipend of one hundred marks per child up to one thousand marks maximum. The following year they began dispensing ten marks per month for the fifth child and any subsequent child under age sixteen.[33]

The Nazi regime also promoted large families through propaganda efforts and by honoring German mothers with a large brood. Hitler, Goebbels, and other Nazi leaders continually assured women that, even though men were the leaders of Nazi society, the role of women was vitally important, too. Nazi propaganda and education lauded women with large families as selfless benefactors of society. Hitler declared in January 1937, "Every mother who has given our *Volk* a child in these four years has contributed, by her pain and her happiness, to the happiness of the entire nation."[34] In 1934 the Nazi regime made Mother's Day a national holiday to honor women. In December 1938 Hitler announced that women bearing many children would receive a medal, the German Mother's Cross. Honoring women with an award for reproducing was an idea that had been floated already before World War I by the eugenicist Gruber.[35] On Mother's Day in 1939 about three million women received their medallions: bronze for four children, silver for six children, and gold for eight or more children. Hitler Youth were instructed to snap to attention and salute women wearing their medals.[36]

Another way the Nazis tried to promote population expansion was by keeping contraception out of the hands of healthy German women. In a particularly sarcastic passage of *Mein Kampf* Hitler pilloried those who supported the use of contraceptives by healthy German women.[37] In the first several months after coming to power, the Nazis closed down birth control clinics and dismantled organizations promoting birth control, incarcerating many of the leaders. They enforced the existing law that banned advertising contraceptives, which the previous regime had largely ignored. In January 1941 Himmler took more drastic measures, banning the production and sale of most contraceptives, because he was concerned that the war would reduce the birthrate further.[38]

Nazi persecution of homosexuals also flowed from Hitler's emphasis on prolific reproduction. Since homosexuality did not contribute to reproduction, Hitler and his regime considered homosexuals useless and retrograde. Already in 1930 the Nazi leader Wilhelm Frick introduced a bill into the German parliament to castrate homosexuals.[39] Himmler was especially zealous about combating homosexuality, establishing an SS Bureau to Combat Homosexuality and Abortion in 1936.[40] The Nazi regime arrested and convicted about 50,000 homosexuals, many of whom were subjected to brutal treatment in concentration camps.[41] In November 1941 Hitler even signed a decree making homosexual offenses among SS members and policemen a capital offense.[42]

Two months earlier Hitler had explained to Goebbels the Darwinian underpinnings of his opposition to homosexuality. After remarking that homosexuality should not be tolerated, especially in the Nazi Party and the army, Hitler continued:

> The homosexual is always disposed to drive the selection of men toward the criminal or at least sickly rather than the useful in the selection of men. If one would give him free rein, the state would in time be an organization of homosexuality, but not an organization of manly selection. A real man would defend himself against this endeavor, because he sees in it an assassination of his own evolutionary possibilities.[43]

When Hitler used the term selection in this conversation, he was using it in the sense of biological selection. He believed that homosexuality led to biological degeneration that would favor hereditary illness and hereditary criminality. His opposition to homosexuality was thus based on a desire to advance the "evolutionary possibilities" of virile heterosexual males.

However, though Hitler did see homosexuality as an aberration that was on the whole harmful to the biological health of the German people, he was flexible in his condemnation of homosexuality. He was well aware of the homosexual tendencies of his friend Ernst Röhm, but he winked at it, because Röhm was an important and influential supporter. According to Hitler's photographer, Heinrich Hofmann, Hitler declared that Röhm's "private life is of no concern of mine as long as the necessary discretion is maintained."[44] In this case Hitler set aside his own moral ideals for pragmatic political considerations. Later,

when Hitler brutally purged Röhm and other SA leaders in the summer of 1934, it was convenient to trumpet their homosexuality before
the German public, to make the blatantly illegal executions seem like
a moral cleansing.

From a reproductive standpoint the Catholic practice of celibacy
was just as objectionable as homosexuality. Hitler did not publicly
denounce celibacy the way he did homosexuality, since he knew it
would not win him popular support. He also did not want to offend
the Catholic Church any more than necessary. Nonetheless, in private
he revealed his disdain for celibacy, which, he thought, robbed the
German people of valuable progeny. In April 1942 he told his entourage that he would make it harder for the Catholic Church to recruit
youth for the priesthood after the war was over. He also threatened to
dissolve all monasteries to free men from their vows of celibacy.[45] In
1940 the Nazi regime issued a decree banning healthy persons from
entering cloisters.[46]

While many of the above policies and plans jibed well with traditional moral standards, other policies marked a sharp break with conservative attitudes. Nazis no longer viewed marriage as a holy union of
two people for better or for worse, in sickness and in health. Rather,
they considered it an institution solely for producing children (and only
healthy ones, to whatever extent possible). Thus, they altered divorce
laws to correlate with these priorities. In 1938 they passed a law allowing
divorce in cases of infertility or if one spouse refused to procreate. One
of the most controversial parts of the law was the permission for divorce
in cases where the couple was separated for more than three years (since
separated couples could not contribute to the desired increase in births).
In cases of divorce for infertility, men were encouraged to marry again
by being partly relieved of alimony payments upon founding a new
family.[47] On several occasions in 1942 Hitler told his associates that he
favored divorce for bad marriages.[48] He clearly did not regard marriage
as a sacrosanct institution.[49]

Hitler and his regime were more interested in increasing the birthrate
than they were in protecting the institution of marriage. Aside from the
dangers of contracting syphilis, Hitler did not seem concerned about
premarital or extramarital sex per se. In a private monologue, Hitler told
associates during the war that soldiers could not be expected to abstain
from sex. "If the German man is prepared as a soldier to die uncon-
ditionally, then he must also have the freedom to love uncondition-
ally.... One cannot come to the soldier with the church's doctrine of
self-denial in the realm of love, if one wants to keep him battle-ready."[50]

Hitler was certainly not interested in upholding Christian prohibitions on fornication, except where it converged with his own goals. Though not all scholars agree about Hitler's own sexual activity, few suppose that he lived a life of sexual abstinence, even though he never married until the final day of his life.

In addition to new divorce regulations, some Nazi agencies encouraged illegitimacy.[51] In Berlin, a major Nazi exhibit, "The Wonder of Life," opened in March 1935. By claiming in one display that "immaculate conception" was any healthy, fit child that was conceived through a loving relationship, it effectively snubbed Catholic values and endorsed fornication.[52] The Aid Organization for Mother and Child, founded in 1934 as a branch of the National Socialist People's Welfare Organization, provided assistance to unwed mothers, as well as married ones. Another important organization promoting illegitimacy was Himmler's *Lebensborn* (literally "Spring of Life"). Lebensborn was created in December 1935 with the approval of Hitler to provide maternal care for pregnant women, especially those carrying babies fathered by SS men. Lebensborn maternity homes provided excellent care for these women, whether they were married or not. For unmarried women, they provided comfortable refuges from disapproving relatives and neighbors. Their primary purpose was not social compassion, but rather improving the German racial stock by encouraging reproduction by those deemed superior.[53]

The advent of war in 1939 gave greater urgency to breaking down taboos against sexual relations outside marriage, since German men were dying, reducing the population. On October 28, 1939, Himmler exhorted his SS men and policemen to reproduce more, whether inside or outside of marriage.[54] His order, approved by Hitler, stated,

> Beyond the boundaries of perhaps otherwise still necessary bourgeois laws and customs it will also outside of marriage be an important responsibility for German women and girls of good blood, not lightly, but rather in profound moral seriousness, to become the mothers of children of soldiers who are going to the front and of whom fate alone knows whether they will return or fall in battle for Germany.[55]

Himmler apparently followed his own advice, fathering two children by a mistress in the 1940s.[56] He promised that the Lebensborn would make sure the wives, girlfriends, and babies of the SS men would receive adequate prenatal and maternity care while the men were away at war.[57] Himmler's views went public in January 1940, when the SS weekly magazine, *Das schwarze Korps*, aroused controversy by publishing an article encouraging women to bear illegitimate children. It chided women, even unmarried women, who shirked their duty to procreate, comparing them to army deserters.[58] Rudolf Hess also encouraged German soldiers to reproduce outside marriage in an open letter to a single woman published in the official Nazi newspaper. He promised that if a soldier died in battle after getting his fiancée pregnant, the child would be considered the soldier's legitimate child.[59]

Hitler completely supported relaxing the taboo on illegitimacy. In February 1934 he told his entourage that his regime would see to it that illegitimate children were put on par with legitimate children, because population expansion was vital.[60] In a monologue in May 1942 he made even clearer that marriage was not sacrosanct. He told his entourage that in areas of the German Reich with poor racial qualities "racially valuable military units" should be sent to "renew the blood of the population" by copulating with locals. Anyone who complains that this will damage the morality of the German people is a hypocrite, he continued. Though he considered the ideal for reproduction a loving, lifelong relationship between a man and woman, nonetheless population growth and racial quality took priority over traditional notions about the sanctity of marriage.[61]

As the bloodletting increased on the Eastern Front, Hitler and Himmler became more concerned about the reduced number of men. They discussed allowing polygamy to help repopulate Germany. Hitler preferred polygamy and illegitimacy to the alternative: some women going without children. "A girl, who has a child and cares for it, is superior to an old spinster," he declared in March 1942.[62] According to Felix Kersten, Himmler's private physician, in 1943 Hitler was considering altering marriage laws after the war to allow war heroes to marry more than one wife. This temporary measure would then be evaluated to determine if monogamy should be retained or not. Himmler was of the opinion that present monogamous rules were immoral and that polygamy would be beneficial.[63]

Despite Nazi attempts to increase the birthrate, the actual achievements were not all that impressive. The birthrate did increase in the first several years of the Nazi regime, but only from an abnormally low figure caused by the Great Depression. Most families in Germany in the 1930s remained small, and the birthrate remained considerably lower than it had been in the early 1920s and earlier. The improving economic conditions in the 1930s probably did more to encourage reproduction than the other specific policies aimed at promoting large families.[64]

Nonetheless, whatever the practical effects, Hitler's concern for fostering population growth flowed from his desire to improve the human species. He believed increasing the birthrate would improve the biological quality of the German people and also allow them to expand at the expense of surrounding "inferior" races. Though his sexual morality and stress on monogamous families often coincided with conservative moral values, his goals were quite different. As we shall see in the following chapter, these differences were also reflected in the sexual morality inherent in Nazi eugenics policies, which also flowed from evolutionary ethics.

CHAPTER SEVEN

Controlling Reproduction to Improve the Human Species

Hitler's sexual morality aimed not only at increasing the German population, but also improving its biological quality. As we have seen, even the drive for higher birthrates was motivated by the belief that it would biologically elevate the German people by begetting more geniuses and superior individuals. The push for higher birthrates, eugenics, and racial purity were all part of the program to biologically reinvigorate the German people. As Walter Gross explained in the foreword to Hitler's pamphlet *Volk und Rasse* (an excerpt from *Mein Kampf*), Nazism was tackling three major manifestations of racial decline: the sinking birthrate, degeneration through hereditary illness, and racial mixture.[1]

These same three elements figure prominently in the booklet, *Why Are We Fighting?*, which Hitler personally endorsed as an instructional tool to inculcate the Nazi worldview into German soldiers. It stressed the centrality of biological improvement for the Nazi worldview and claimed that one of the main goals of Nazism was producing a "new human type" through "the preservation of purity of our blood and the higher evolution of our blood." This booklet repeatedly emphasized the need not only to maintain the present level of the German race, but also to foster evolutionary progress: "Thus the main demand of National Socialism is not only to preserve the racial hereditary substance of the German *Volk*, but also to increase its value." It called for biological improvement, using not only the language of breeding, but also explicitly evolutionary language, stating, "National Socialism strives for the higher evolution of the *Volk*." This drive to move the German race to a higher evolutionary plane was a moral imperative for

Nazis, as is evident in the following statement: "Our racial idea is only the 'expression of a worldview,' which recognizes in the higher evolution of humans a divine command." When discussing how to drive evolution forward, the booklet discusses both population expansion and eugenics measures. It thus made explicit what was always implicit in Nazi eugenics propaganda and policies: The purpose of improving hereditary health was to advance humanity in the evolutionary process.[2]

For Hitler hereditary health was one of the chief virtues, so promoting hereditary health was one of the most important tasks the state could perform. He recognized that artificial selection by the state was not the same as natural selection, but he nonetheless believed that promoting the strong, healthy, and intelligent at the expense of the weak and ill was consistent with the laws of nature. Artificial selection—generally called eugenics or race hygiene by its proponents—would counteract the supposedly harmful influences of modern institutions that aided the weak and sick. It would counteract the elimination of natural selection, which he blamed for producing biological degeneration.

Hitler's quest for improving the hereditary health of the German people involved both positive and negative eugenics. Positive eugenics focused on measures to encourage the more prolific reproduction of those with "good" heredity, while negative eugenics promoted policies to hinder the reproduction of those deemed hereditarily "unfit" or "inferior." Most of the Nazi pro-natalist policies I have discussed in the previous chapter were examples of positive eugenics, because they only encouraged reproduction of healthy Germans. For example, marriage loans and child benefits were not available to those suffering from congenital illnesses, but were only granted to healthy German couples. Lebensborn assisted only those considered biologically superior. Thus Nazi pro-natalist policies were discriminatory, applicable only to those deemed hereditarily fit.

Under Nazism negative eugenics targeted two allegedly inferior groups: those defined as racially inferior and the disabled. Hitler's pro-natalism was certainly not intended for them. In the first year of his rule, Hitler and his regime began introducing measures to hinder the disabled from reproducing. Before World War II they did not take any similar measures to prohibit Jews or other races in Germany from reproducing among themselves. However, in 1938 the Nazi regime told Jews that the abortion laws did not apply to them.[3] Furthermore, the Nazi regime introduced measures to hinder racial mixing.

Banning Racial Mixing

We have already seen the importance of racial inequality and racial struggle in Hitler's worldview. His racism influenced his sexual morality, because he considered racial mixing dangerous. Hitler insisted that miscegenation was one of the most important factors leading to biological decline, thus hindering further evolution. He considered this shameful and dangerous, warning in *Mein Kampf*, "*The sin against blood and the race are the original sin in this world and the end of a humanity which surrenders to it.*" The sin Hitler was preaching against here was racial mixing, which was so evil that he considered it the "original sin" (as Lanz von Liebenfels had earlier). The effects he prophesied were nothing short of catastrophic—the end of humanity and culture.[4]

The first point that Hitler made in the chapter on "Race and Nation" in *Mein Kampf* was that nature teaches "the inner segregation of the species of all living beings on this earth." The problem with crossing organisms with differing value is that the offspring will have traits that lie somewhere between the higher and lower parent. Hitler continued,

> Consequently, it [the offspring] will later succumb in the struggle against the higher level. Such mating is contrary to the will of Nature for a higher breeding of all life. The precondition for this does not lie in associating superior and inferior, but in the total victory of the former. The stronger must dominate and not blend with the weaker, thus sacrificing his own greatness. Only the born weakling can view this as cruel, but he after all is only a weak and limited man; for if this law did not prevail, any conceivable higher evolution of organic living beings would be unthinkable.[5]

Hitler then moved seamlessly (and illogically) from species to human races, implying that races are subject to the same natural segregation that species are. Racial crossing, he claimed, leads to lowering the level of the higher race, both physically and intellectually. Racial mixing, then, is "nothing else but to sin against the will of the eternal creator" and "to rebel against the iron logic of Nature."[6] Thus one of the highest commands in Hitler's moral code was: Thou shalt not have interracial sexual relations.

The reason for this prohibition against interracial procreation was because in Hitler's opinion it would hinder or even reverse the evolutionary process. After discussing the benefits of the struggle for existence on biological improvement in *Mein Kampf*, Hitler asserted, "No more than Nature desires the mating of weaker with stronger individuals, even less does she desire the blending of a higher with a lower race, since, if she did, her whole work of higher breeding, over perhaps hundreds of thousands of years might be ruined with one blow."[7] Indeed Hitler not only believed theoretically that racial mixture could lead to decline, he thought that historically it had done so repeatedly.

If the Aryan race was the progenitor of all earlier civilizations, as Hitler thought it was, what had caused the collapse of these civilizations? For Hitler the answer was simple: Racial mixing brought about their decline and ultimately their downfall. "All great cultures of the past perished only because the originally creative race died out from blood poisoning." By blood poisoning Hitler meant racial mixing, for he explained that in earlier cultures the Aryan lords had after a few centuries begun to mix with their subjects. They thereby lost their superior racial traits, leading to biological and cultural decadence. Political decline simply mirrored their biological condition. One example that Hitler provided in *Mein Kampf* to illustrate the perils of racial mixing was the Americas. He thought the Germanic peoples had not mixed much with other races in North America, making the United States powerful technologically and militarily. However, Latin America's weakness resulted from the Spaniards intermingling with the Indians.[8]

Germany's distress in World War I and its aftermath also flowed from racial decline preceding the war, according to Hitler. "The deepest and ultimate reason for the decline of the old Reich lay in its failure to recognize the racial problem and its importance for the historical development of peoples," he asserted. "All really significant symptoms of decay of the pre-War period can in the last analysis be reduced to racial causes." Decades-long racial decline caused the disastrous military defeat in World War I. No reforms could ultimately rescue Germany, unless it restored racial purity.[9]

Indeed Hitler promised regeneration for Germany through purification of the German race. He explained that racial half-breeds are not only inferior to the higher parent, but "they lack also the unity of will-power and determination to live." Even the vaunted willpower so important to Hitler depended on one's racial character. Hitler thought racial regeneration was possible as long as enough superior racial stock was still present. He obviously had faith that Germans still had sufficient Aryan blood to initiate this racial purification. While obeying racial instincts would be the most important factor driving this process, the state would also intervene. After alleging that racial mixture would completely destroy culture and undermine the "mission of humanity," he warned, "Anyone who does not want the earth to move toward this condition must convert himself to the conception that it is the function above all of the Germanic states first and foremost to call a fundamental halt to any further bastardization."[10] He continued,

> The generation of our present notorious weaklings will obviously cry out against this, and moan and complain about assaults on the holiest human rights. *No, there is only one holiest human right, and this right is at the same time the holiest obligation, to wit: to see to it that the blood is preserved pure and, by preserving the best humanity, to create the possibility of a nobler evolution of these beings.*[11]

Hitler thus tied together his imperative to maintain racial purity with the goal of fostering evolution.

The notion that racial mixing leads to biological and cultural decline was widespread among racial thinkers in the early twentieth century. Gobineau was the most famous exponent of this view in the nineteenth century, promoting it in his influential book, *Essay on the Inequality of Human Races* (1853–1855). Gobineau's opposition to racial mixing spread widely in German society in the early twentieth century through the Gobineau Society under Ludwig Schemann. Another important advocate of Gobineau's ideas was the racial theorist Ludwig Woltmann and his circle, who also warned about the perils of miscegenation. Their message about the threat of racial mixing was incorporated into academic anthropology by Eugen Fischer, professor at the University of Freiburg. As a young scholar, Fischer, already a devotee of Woltmann, travelled to German Southwest Africa to investigate miscegenation firsthand. His book, *The Rehoboth Bastards and the Bastardization Problem among Humans* (1913), examined a community of descendants of European men and African women. Fischer believed that racial crossing usually produces progeny approximately midway between the races of the parents. Thus he opposed racial mixture and supported racial segregation in German colonies.[12] Fischer's work was important in giving a scientific patina to opposition to miscegenation.

Hitler probably never read Fischer's book, but many anthropologists and racial thinkers were promoting the same or similar ideas by the 1910s and 1920s, so he could have picked up the idea from any number of sources. One possible influence in this regard was Lanz von Liebenfels, the Aryan supremacist thinker in Vienna who reported that

Hitler read his periodical *Ostara* in Vienna when he lived there. Though we are not sure if Lanz von Lebenfels testimony is accurate, some of Hitler's terminology about racial mixing does seem to derive from *Ostara*. Another likely influence was Günther. Even though Günther acknowledged that no races are really pure, he still portrayed racial mixture as usually deleterious. He specifically criticized intermarriage between Germans and Jews, calling this *Rassenschande* (racial disgrace), a term the Nazis would use liberally later.[13]

Since the most influential opponent of racial mixing, Gobineau, published his famous book on racism before Darwin published his theory, it is quite obvious that opposition to racial mixing did not derive from Darwinism. However, as we have already seen (chapter 3), many of Gobineau's disciples in the 1890s and thereafter (Schemann, Woltmann, Fischer, and others) integrated his ideas into a Darwinian framework. By the 1890s the earlier triumphant optimism of the inevitability of evolutionary progress had given way to widespread fears that the evolutionary process might not always lead upward. Evolutionists warned about the specter of biological degeneration. Thus Gobineau's followers interpreted his ideas about biological decline through racial mixing as a part of the evolutionary process. Hitler certainly interpreted racial mixing in this way.

Indeed Hitler's regime translated his concern about the immorality of racial mixture into policy during the Third Reich. Already in 1923 Hitler had indicated that the Nazi Party supported a ban on intermarriage between races (undoubtedly meaning primarily Germans and Jews), because the offspring of mixed races lack vitality and are thus "a valueless product."[14] Long before coming to power the Nazi Party enforced a ban against members marrying Jews or other non-Aryan races, and the SS had stringent requirements for proving Aryan ancestry. Once the Nazis seized power, they began expanding the prohibitions against miscegenation. On June 30, 1933, they passed legislation forbidding any government official from marrying non-Aryans.[15]

Hitler announced more sweeping Nazi legislation prohibiting interracial marriage and sexual relations at the Nuremberg Party Congress in September 1935. One of these Nuremberg Laws forbade Jews from marrying those "of German or kindred blood." However, the law had not been properly vetted and prepared beforehand, so for about two months thereafter Nazi officials discussed and debated how to define a Jew and how to apply the law to those who were "half-Jews." Many scholars have noted that the Nazis were singularly unable to provide any biological definition of a Jew, so they were forced to define Jews based on their religious identification. Nazis defined as a Jew anyone having three or four grandparents who practiced the Jewish religion. Synagogue records, not biological markers, were decisive. Still, this does not prove that religion was more important to the Nazis than race. The opposite is true, and that is why one's fate was determined by one's grandparents, not one's parents or one's own religious affiliation. Also, in September 1933 the First Supplementary Decree to the Civil Service Law clearly stated that "it is not religion which is decisive but rather *descent, race, blood.*" The Supplementary Decree further stated that even if the parents or grandparents did not belong to the Jewish religion, if Jewish ancestry could be established some other way, the Civil Service Law could still apply to that person.[16] Having to use religion to measure race was inconsistent and showed the poverty of Nazi racial theory, but the Nazis considered it an unfortunate stopgap measure while scientists searched for a more reliable biological marker.[17]

Even more importantly, the debates within Nazi circles over what to do about those who were half-Jews or quarter-Jews (called *Mischlinge*) shows that biology really was important in framing Nazi racial laws. One of the key disagreements among Nazi racial experts was about how to apply Mendelian genetics to interracial marriage and sexual relations. On September 25, 1935, one of the leading race experts of the Interior Ministry, Arthur Gütt, wrote a brief synopsis giving his perspective on the issue. He argued that Mendelian genetics made the problem of the *Mischlinge* almost insoluble. Nonetheless, he supported allowing quarter-Jews to marry Germans, and he thought that half-Jews should be sorted by anthropologists to determine their racial fitness. If they were deemed to have sufficient Germanic characteristics, then they could marry Germans, but they should not marry other *Mischlinge*. However, Karl Astel, a prominent Nazi racial scientist at the University of Jena, disagreed with Gütt. On October 8 he submitted a rebuttal to Himmler, arguing that because of Mendelian genetics, no *Mischlinge* should be allowed to reproduce with Germans. Otherwise Jewish hereditary traits could resurface in subsequent generations, even if they were latent presently.[18]

Hitler was actively involved in the discussions within his regime about how to define the Nuremberg Laws. He intervened five times in the framing of the supplementary decrees, which provided the official interpretation of the laws.[19] In the First Supplementary Decree to the Nuremberg Laws (November 14, 1935), Gütt's position on the *Mischlinge* prevailed, since it allowed quarter-Jews to count as

Germans for almost all legal purposes. For instance, quarter-Jews, just like Germans, were not allowed to marry or have sexual relations with Jews. However, there was one way that these quarter-Jews were not treated as full Germans: they were not allowed to marry among themselves. The decree also forbade marriage between half-Jews and Germans, unless special permission was obtained (and it was seldom granted). Mendelian genetics did not provide a conclusive answer to the problem of racial mixing, so tactical considerations—especially the problem of offending German relatives of *Mischlinge*—influenced Nazi policy. However, scientific considerations were important and played an important role in the debate over racial policy.[20] According to the official Nazi commentary on the Nuremberg Laws, the law was framed in such a way that it would eventually lead to the elimination of the mixed race.[21]

Hitler apparently followed this argument over Mendelian genetics. He already knew about the significance of Mendelian genetics for racial mixture at least by 1928, for he mentioned in his *Second Book* that because of Mendelian genetics some offspring in a racially mixed marriage would favor one race, while some siblings might favor the other. Thus within one family the racial qualities would be uneven.[22] Cornelia Essner claims that a speech Hitler gave in late September 1935 showed "surprisingly good racial-biological knowledge," probably in part because he read Gütt's policy paper.[23] However, Hitler apparently wavered between Gütt's and Astel's positions on the application of Mendelian genetics to racial mixing. In a monologue in December 1941 Hitler claimed that even though Jewish racial characters show up in those of mixed ancestry in the second or third generation, those traits usually vanished by the seventh, eighth or ninth generations. He stated, "The Jewish character is sorted out through the Mendelian law, evidently restoring purity of blood."[24] However, five months later Hitler reversed himself, expressing regret that he allowed so many half-Jews in the military. In this monologue he claimed that because of Mendelian laws even after four or five or six generations a "pure Jew" could emerge. These "pure Jews" emerging generations after racial mixture occurred "constitute a great danger," Hitler told his associates. Two months later Hitler discussed this theme again, claiming that a certain man manifested Jewish characteristics, even though his last Jewish ancestor had been born in 1616. This confirms, Hitler reported, that "in the course of generations a racially pure Jew can emerge by Mendelian laws." This proved to Hitler that *Mischlinge* should not be accorded equal status to full-blooded Germans.[25] Thus it seems that Hitler's stance toward the *Mischlinge* became harsher during the war. Though many half-Jews and almost all quarter-Jews would survive the Nazi Holocaust because they were not treated as full-Jews, James Tent is probably right to suggest that they would have eventually been victims had the Nazi regime lasted longer.[26]

However, Hitler's ideology and Nazi policy aimed not only at eliminating mixture with Jews, but also with other races deemed inferior.[27] In 1930 the Nazi Party introduced a bill into parliament to forbid racial mixing with Jews and with "colored races."[28] Nazi opposition to racial mixture with non-Jewish "inferior" races was also apparent from the discussions about interracial marriage within Nazi government agencies, leading up to the Nuremberg Laws. In April 1935 the Main Office for the *Volk*'s Health held a meeting to draft legislation concerning racial criteria for citizenship and inter-racial marriage. Leading Nazis concerned about racial policy were present, including Himmler, Walther Darré, Ernst Rüdin, Julius Streicher, Walter Gross, Gütt, and Gerhard Wagner. They formu-lated four racial categories: (1) German and related (*deutschstämmig*); (2) neighboring races; (3) foreign races (*fremdstämmig*); and (4) Jews. The draft legislation they produced allowed intermarriage between Germans and neighboring races, but not between Germans and for-eign races and not between Germans and Jews.[29] Even more sig-nificantly, the First Supplementary Decree to the Nuremberg Law expanded the marriage prohibition to include more than just Jews. It stated, "Further, a marriage shall not be contracted, if a progeny that endangers the preservation of German blood can be expected to issue from it." The official Nazi commentary on this law made clear that this included blacks, Gypsies, and most non-European peoples.[30] Though Jews were the primary target of the Nuremberg Laws, the supplementary decrees and official commentary expanded racial dis-crimination to other races the Nazis considered inferior.

Hitler's contempt for black Africans led him to strenuously oppose miscegenation between Germans and blacks. After World War I, many racists, including Nazis, fulminated against France for bringing black African troops into the Rhineland. False rumors about German women being raped by these black troops circulated widely in Germany in the 1920s. The truth was more prosaic: hundreds of German women had consensual sexual relations with black soldiers, begetting the so-called "Rhineland bastards." Hitler bizarrely claimed in *Mein Kampf* that the "contamination by Negro blood on the Rhine" was part of the Jewish conspiracy "to begin bastardizing the European continent at its core and to deprive the white race of the foundations for a sovereign existence through infection with lower humanity."[31]

Hitler's outrage about miscegenation with black Africans was shared by many leading eugenicists, including Rüdin, who chaired the Interior Ministry's committee on racial policy that discussed the sterilization of the "Rhineland bastards" in 1935. When the Nazi regime finally secretly (and illegally) sterilized several hundred "Rhineland bastards" in 1937 to keep them from passing on their "inferior" racial traits, Eugen Fischer and other leading anthropologists cooperated by helping to identify the target population.[32] Nazi opposition to miscegenation with blacks also led them to put black American athletes in Berlin for the 1936 Olympics under surveillance, lest they consort with German women. German women who came inappropriately close to the blacks were warned by Nazi authorities to keep their distance.[33]

Hitler also opposed racial mixture with Slavs (except with those deemed racially on par with Germans). In April 1940 Hitler (through Bormann) instructed his Interior Ministry to dismiss any government official who had sexual relations with Poles or Czechs.[34] After that time all Germans wanting to marry Czechs had to get permission from Nazi authorities.[35] Ten days after invading Poland in September 1939 Hitler told Himmler that if any Polish POWs were caught having sexual relations with German women, the man would be shot, while the woman would be publicly pilloried and sent to a concentration camp.[36] In October 1940 Hitler again warned some of his closest associates about the perils of allowing racial mixture between Poles and Germans.[37] In February 1942 Hitler issued a decree forbidding German soldiers from having relations with Polish women.[38] Any Polish woman caught having sexual relations with a German would be committed to a brothel.[39]

When Germany began importing millions of Slavic slave laborers in the early 1940s, the Nazi regime did everything possible to prevent interracial sexual relations. In most cases they either sent Slavic women along with the men or else established brothels with Slavic women for them.[40] They issued strict warnings to both Germans and the Slavic workers not to engage in sexual relations with each other. Starting in February 1940 all Polish laborers in Germany had to wear a symbol marking their pariah status. Every German farmer employing Slavic workers received a notice, stating, "Keep German blood pure! That holds for men as well as for women! Just as it is the greatest shame to have sexual relations with a Jew, so every German who has intimate relations with a Polish man or woman transgresses." German women who were caught having sexual relations with Slavic laborers were usually pilloried and then sent to a concentration camp. The Slavic male offender was executed.[41]

Ironically, the Nazis did not automatically reject as racially inferior all progeny produced by German-Slav miscegenation. Hitler decreed in October 1943 that children of German men and native women in the Eastern occupied zones would be cared for by the German state, as long as they were deemed "racially valuable."[42] These children were often taken from their mothers and sent to Germany to be raised by German parents or in German institutions. Likewise, if foreign women workers became pregnant while in Germany, Himmler directed that children deemed "good racially" would be raised in special homes in Germany, while those considered "bad racially" would be sent to separate institutions.[43]

Nazi concern for racial purity led to numerous marriage restrictions for various segments of German society. Nazi Party members, for instance, were not allowed to marry Jews or blacks, and they could only marry Czechs, Poles, or Magyars with the approval of Nazi officials (*Gauleiter*). SS members had to get permission from their superiors to marry, and their fiancées were vetted for racial and hereditary health. From 1936 on, German soldiers needed permission to marry, which was only given if their fiancée was of German or related ancestry.[44] In 1940 Hitler decreed that no member of the Foreign Service could marry a non-German without permission from the Foreign Minister, and no one married to a non-German could be hired by the Foreign Service.[45] Two years later he told colleagues that he regularly denied soldiers' applications for marriage with foreigners.[46] However, as Hitler made clear on numerous occasions, marriages between German soldiers and Scandinavian or Dutch women were perfectly acceptable, since he considered them fellow Aryans.[47]

After World War II began, the Nazi regime introduced a variety of measures to discourage reproduction among the allegedly inferior peoples in their occupied territories in Poland and the Soviet Union. In the part of Western Poland annexed as the Wartheland, the regime raised the minimum age for marriage to twenty-eight for Polish males and twenty-five for Polish females.[48] In July 1942 Hitler railed at the "idiot" who suggested banning abortions in occupied Eastern territories. Rather, he stated, the German occupiers should encourage abortion and contraception, and they should refrain from providing any medical care to the native populations.[49]

Eugenics

Hitler's drive to guide biological evolution forward made him a fanatical supporter of eugenics—also known as race hygiene in Germany—by at least 1923. Most German eugenicists, including its founding fathers August Forel, Alfred Ploetz, and Wilhelm Schallmayer, considered their program an application of evolutionary principles to ethics and society. Ploetz informed a friend in 1892 that his main ideas about eugenics were drawn from Darwinism, and he often praised Haeckel as a formative influence on his world view.[50] He also recruited the two leading Darwinists in Germany—Haeckel and Weismann—to became honorary members of the Society for Race Hygiene when he founded it in 1905. Schallmayer's most influential book on eugenics, *Heredity and Selection* (1903), was the winning entry in the Krupp Prize Competition. It responded to the question, "What do we learn from the principles of biological evolution in regard to domestic political developments and legislation of states?"[51] Schallmayer confided to another leading eugenicist that eugenics was indissolubly bound together with Darwinian theory.[52] Almost all early eugenicists—both inside and outside Germany—agreed with Ploetz and Schallmayer. An illustration used at the Second International Eugenics Congress in 1921 in New York City announced, "Eugenics is the self direction of human evolution."

Hitler's eugenics ideology was clearly grafted onto his vision of biological determinism, Darwinian struggle, and evolutionary advance that played such a prominent role in his thinking from his earliest speeches. However, of all the major elements of Hitler's worldview, eugenics is conspicuously absent from the Nazi Twenty-Five Point Program of 1920. Indeed, Hitler did not explicitly endorse eugenics until 1923. However, in one of his earliest speeches in August 1920 he did clearly embrace biological determinism, an idea underpinning eugenics ideology. In that speech he claimed that the harsh climate during the Ice Age had contributed to the health and vitality of the Nordic race by selecting out the best: "Whoever was weak or sick could not survive this frightful period, but rather sank prematurely into the grave, leaving a generation of giants in strength and health....all inferior and weak individuals gradually died out of this race, leaving only the healthiest bodies."[53] His paean to natural selection stopped short of endorsing artificial selection, but it shows that Hitler by that time already embraced many of the presuppositions that drove the eugenics movement.

Figure 7.1 Certificate from the Second International Eugenics Congress, 1921.

In an interview with an American journalist in October 1923 Hitler enthusiastically endorsed eugenics. The journalist reported: "Hitler believes in eugenics. In order, he says, to make our people worthy of the crown of citizenship, we must cut out every cancer that corrodes our life. Syphilitics, alcoholics, must be isolated. They must not be permitted to reproduce." Hitler then contemptuously dismissed the "false humanitarianism that teaches us to preserve the unfit," calling this "diabolically cruel." He argued that not only physical and mental disabilities, but also crime was the product of bad heredity. Using inflammatory language, he asserted that in his future state "there will be no room for the alien, no use for the criminal, no use for the diseased, no use for the wastrel, for the usurer or speculator, or anyone incapable of productive work." Despite the innuendo about getting rid of such persons, Hitler's only concrete proposal was to isolate such persons to prevent them from reproducing.[54]

In *Mein Kampf* Hitler vigorously supported eugenics, stridently calling for an end to biological degeneration brought on by allegedly misguided humanitarianism. In an early section on the lessons he supposedly learned from his "struggle for existence in Vienna," he outlined a twofold path for solving social problems: (1) creating "better foundations for our evolution"; and (2) "brutal determination in breaking down incurable tumors." He then critiqued humanitarian efforts to solve social problems, which violated the laws of nature. He stated:

> Just as Nature does not concentrate her greatest attention in preserving what exists, but in breeding offspring to carry on the species, likewise, in human life, it is less important artificially to alleviate existing evil, which, in view of human nature, is ninety-nine per cent impossible, than to ensure from the start healthier channels for future evolution.

Hitler then claimed that the way to solve social problems is to get rid of any policies or institutions that cause biological degeneration. He specifically mentioned criminals as one manifestation of degeneration. Though he did not propose any specific eugenics measures in this passage, he made clear his desire "brutally and ruthlessly to prune off the wild shoots and tear out the weeds."[55]

In the second volume of *Mein Kampf* Hitler devoted several pages, including a long passage in italics, to eugenics. He argued that the state *"must see to it that only the healthy beget children . . . It must declare unfit for propagation all who are in any way visibly sick or who have inherited a disease and can therefore pass it on."* One of the chief tasks for the state would be to educate its citizenry on the importance of eugenics. It must teach people that it is a crime and dishonor to propagate bad heredity. Hitler promised that preventing the hereditarily sick from procreating over several centuries would "lead to a recovery which today seems scarcely conceivable." Aside from education and voluntary renunciation, Hitler never specified in this book what measures the government should take to ensure that those with hereditary illnesses do not procreate.[56] However, in another place in *Mein Kampf* he did briefly mention that after completing military service men should receive a health certificate "confirming his physical health for marriage," which seems to imply that he favored requiring health certificates for those wanting to marry.[57]

While we do not know exactly what eugenics literature Hitler read to form his early views, it seems likely that it was mediated by his friend, the medical publisher Julius Friedrich Lehmann. Lehmann published many works on eugenics, including the influential two-volume text on human genetics and eugenics by Erwin Baur, Eugen Fischer, and Fritz

Tafel 12.

Die Drohung des Untermenschen.

Es treffen auf:

Männliche Verbrecher: 4,9 Kinder   Eine kriminelle Ehe: 4,4 Kinder

Eltern von Hilfsschulkindern: 3,5 Kinder

Die deutsche Familie: 2,2 Kinder   Akademikerehe: 1,9 Kinder

Figure 7.2 "The Threat of the Subhuman," showing fertility rates of criminals (top two figures), parents with special education children (middle figure), the average German family (bottom left), and the German academic (bottom right). (from Volk in Gefahr)

Lenz. In 1931 Lenz, who wrote most of the material on eugenics in this work, boasted that Hitler had read the second edition of this text while incarcerated in Landsberg Prison. He bragged, "Many passages from it are reflected in Hitler's turn of phrase [in *Mein Kampf*]. In any case he appropriated for himself the essential ideas of race hygiene and its importance with great mental responsiveness and energy."[58] Though we do not know for sure, it is highly likely that Lehmann sent Hitler a copy of the second edition of the Baur-Fischer-Lenz text when it was published, since in the 1920s Lehmann regularly supplied Hitler with books he published on racism and eugenics. We know that Lehmann sent him the third edition of Baur-Fischer-Lenz (1927–1931).[59]

Even if Hitler did read the Baur-Fischer-Lenz book in Landsberg Prison, as seems likely, this does not explain how he came to subscribe to eugenics before the Beer Hall Putsch. It is possible that he read the Baur-Fischer-Lenz book before October 1923, since the first edition had been available since 1921. However, it is even more likely that Hitler imbibed eugenics through articles in Lehmann's journal, *Germany's Renewal*. In 1918 and thereafter this journal carried several major articles about eugenics, such as the one by Gruber, "Race Hygiene, the Most Important Task of *Völkisch* Domestic Policy." Hitler could also have learned about eugenics from Günther's discussion of it in *Racial Science of the German Volk*.[60] Lehmann presented Hitler at least three different editions of Günther's famous book, including the 1923 third edition.[61]

Eugenics was not a major theme in most of Hitler's speeches in the 1920s, but it did play a prominent role in his speech to the Nuremberg Party Congress in August 1929. In that speech he explained that the source of Germany's problems was not economic, but rather that it was located in its biological substance. He emphasized human inequality, both of individuals and of races. In order to avoid biological degeneration and improve the human species, the government should implement laws that restrain the "inferior" individuals from reproducing. He wanted to supplement "the natural process of selection" with artificial selection, that is, with eugenics measures.[62]

On July 14, 1933, the same day that the Nazis celebrated the destruction of their political opponents by declaring themselves the only legal political party in Germany, they also passed their first piece of eugenics legislation: the Law for the Prevention of Hereditarily Diseased Offspring. The Interior Minister Wilhelm Frick convened an Expert Committee on Population and Racial Policy in late June 1933; their first order of business was to finalize this legislation. The committee included some of the leading eugenicists in Germany—Alfred Ploetz, Fritz Lenz, and Ernst Rüdin—as well as leading Nazi officials.[63]

This law allowed physicians and directors of hospitals and other institutions to compulsorily sterilize those having specified hereditary illnesses. The list of ailments for which sterilization was permitted included five psychiatric conditions (including the elastic category of feeble-mindedness), three physical conditions, and chronic alcoholism. It set up Hereditary Health Courts to decide on recommendations submitted by physicians. Many leading anthropologists and physicians participated in this process either by sitting on the Hereditary Health Courts or by submitting expert opinions to the courts.[64] Though many states in the United States and a few other European countries had compulsory sterilization laws for the hereditarily disabled, none were so vigorously and fanatically implemented as the Nazi program. In less than twelve years German physicians sterilized about 400,000 people.[65]

In the official commentary on the sterilization law, Gütt, Rüdin, and Falk Ruttke explained the ideological foundations of eugenics. They credited Darwin, Mendel, and Galton with the initial thrust, which was followed up by Ploetz, Schallmayer, Baur, Rüdin, Lenz, and

Figure 7.3 "Sterilization: Not Punishment—but Liberation" (from Nazi periodical)

others. They carefully explained that natural selection was a beneficial process that brought biological progress by causing "inferior" people to die, while the healthy and strong reproduced. The problem for modern society, however, was that with "inferior" people today "the reduced adaptation, as Darwin expressed it, does not lead to eradication, but rather the effect of natural selection has been transformed through civilization into its opposite and thus to *contraselection*."[66] The specter of contraselection, that is, biological degeneration caused by allowing the disabled to reproduce, was ever-present in Nazi eugenics propaganda.

In the discussion of the sterilization law in Hitler's cabinet meeting in July 1933, Hitler personally approved of the legislation. He suggested that habitual criminals be sterilized, too. The Justice Ministry opposed this and Hitler relented, though he requested that forthcoming penal reform would include provisions for sterilizing habitual criminals. The Justice Ministry eventually balked on this, so Gütt, Rüdin, and Ruttke in their official commentary on the sterilization law suggested that many habitual criminals were also feebleminded, and they encouraged physicians to apply the sterilization law to them if possible.[67]

In March 1934 a Hereditary Health Court expanded the sterilization law further by ruling that pregnant women who fell under the purview of the sterilization law could have abortions.[68] In September 1934 Gerhard Wagner gained permission from Hitler to allow abortions for those subject to the sterilization law. He sent a confidential memo to health officials informing them that Hitler would ensure that physicians performing eugenic abortions would not be prosecuted.[69] This decision was enshrined in law in an amendment to the sterilization law in June 1935.[70] Though the law stipulated that eugenic abortions should only be performed with consent of the pregnant woman, unless she was unable to express consent, many women were forced to have abortions.[71] We do not know how many eugenic abortions the Nazis performed, but the number was probably not negligible. Horst Biesold reports that of the 662 deaf women compulsorily sterilized by the Nazis, 57 reported that they had been forced to have abortions.[72]

When Hitler celebrated the first anniversary of his appointment as chancellor on January 30, 1934, he boasted about the sterilization law his regime had passed. He told his parliament that now that most political opponents had been cleared away, only two categories of people dangerous to the state remained: opportunists and "the army of those who were born into the negative side of the racial [*völkisch*] life due to their hereditary predisposition." He called for the state to take "genuinely revolutionary measures" to deal with the hereditarily disabled, though he did not specify what these measures might be. He called the sterilization law only "an initial offensive against this threat," implying that more eugenics measures would follow, as indeed they did.[73]

Hitler also noted in his January 1934 speech that some segments of the German public—especially the churches—were critical of the sterilization legislation. Probably because of this public opposition, Hitler refrained from saying much about eugenics legislation or sterilization after his January 1934 speech. He even instructed Goebbels in June 1935 to suppress publicity about the sterilization law for "sociopsychological reasons."[74] However, various Nazi agencies, most prominently the Racial Policy Office, the National Socialist Physicians' League, and the Health Office in the Interior Ministry, continued with a propaganda offensive promoting eugenics. They organized seminars and lectures on eugenics for schools, medical students, and various Nazi organizations. Walter Gross's Racial Policy Office produced five documentary films between 1935 and 1937 promoting eugenics. The titles of these films—one was named "All Life Is Struggle"—"referred to the social Darwinian ideology of the continuous struggle for survival in human society, hereditary health and race hygiene," according to Ulf Schmidt.[75]

In 1937 Hitler ordered the production of a feature film, "Victims of the Past," to educate the German public on the dangers of hereditary illnesses and the necessity of eugenics policies. As Michael Burleigh has noted, this and the other eugenics films were laced with social Darwinist rhetoric.[76] The narrator of "Victims of the Past" stated,

> Everything weak unfailingly perishes in nature. We have sinned terribly against this law of natural selection in the last decades. We have not only preserved the life [of the weak], but we have even allowed them to reproduce. All this misery could have been prevented, if we had previously prevented the reproduction of the hereditarily ill.[77]

This encapsulates the social Darwinist vision that motivated Hitler and his minions to promote and implement eugenics policies.

After the sterilization law, the next major piece of eugenics legislation aimed at the disabled was the Law for the Protection of Hereditary Health, often referred to as the Marriage Health Law (October 18, 1935). This law came right on the heels of the Nuremberg Laws, and according to Gütt, who was involved in deliberations on these laws, the marriage restrictions in both laws were integrally related. Both were intended to improve the biological health of the German people.[78] The official Nazi commentary on the Nuremberg Laws also contained the commentary on the Marriage Health Law. It claimed that both laws were part of a single package to protect the hereditary health of the German people.[79]

The Marriage Health Law forbade marriage for individuals who had hereditary illnesses already listed in the Law for the Prevention of Hereditarily Diseased Offspring. Though the law required everyone to get a health certificate before marrying, the German health system was not adequate to the task, so local officials only required health certificates if they suspected that one of the parties getting married had a hereditary problem.[80] The First Supplementary Decree to the Nuremberg Laws also required that these health certificates verify that prospective brides and grooms were not from inferior races, again showing the link between the Nuremberg Laws and the Marriage Health Law.[81] Although the Marriage Health Law allowed those who had already been sterilized to marry, in 1936 a German court ruled that a hereditarily ill person could not marry a hereditarily healthy person, even if the former were already sterilized.[82]

Hitler also made clear his support for eugenics by honoring leading eugenicists. In January 1936 he personally granted Ploetz the honorary title of professor for his contributions in organizing and leading the eugenics movement.[83] Hitler bestowed on Rüdin, the leader of the German eugenics organization, the Goethe Medal for Art and Science, one of the highest honors scholars could receive.[84] Hitler also wrote a letter of gratitude to a prominent eugenicist, Philalethes Kuhn, when he retired from his professorship.[85] Before the Nazi period, only the University of Munich had a professorship in race hygiene (though many medical professors in other fields supported eugenics). The Nazi regime established chairs in race hygiene at twenty other universities.[86] The Nazi government promoted leading eugenicists into professorships in other fields, too, such as anthropology and human genetics.

The Nazi regime often ran into pragmatic problems implementing policies based on their ideology, because sometimes one part of their program might conflict with another part. For instance, pro-natalist and eugenics policies sometimes conflicted with each other. Just before Hitler launched war against Poland in 1939, the Nazi regime revoked its requirement for health exams before marriage. The purpose was to allow conscripts to get married quickly, so they would beget children before being sent off to war. However, some Germans who could not pass the health exams likely took advantage of this opportunity to marry. Pro-natalism and eugenics also came into conflict in Nazi policy toward contraceptives. When Himmler banned most contraceptives in 1941, he exempted condoms, because the army considered them essential in combating syphilis.[87] Eugenicists heartily approved of preventing syphilis.

Another reason that Nazi policies did not always correspond to Nazi ideology is because leading Nazis were not always sure what concrete policies would benefit the German race. Nonetheless, Hitler's sexual morality and Nazi policies based on it always aimed at improving the German race and the human species. Pro-natalism, prohibitions against miscegenation, and eugenics were part of a coordinated program to improve the German people biologically. At the Nuremberg Party Congress in September 1937, Hitler bragged about the racial and eugenics policies his regime had pursued. He stressed the sweeping significance of these policies, stating that "the greatest revolution Germany has undergone was that of the purification of the *Volk* and of race hygiene, which was launched systematically in this country for the first time ever." He continued, "The consequences of this German racial policy will be more significant for the future of our *Volk* than the effects of all the other laws together. For they are what is creating the new man." He then invited his audience to look around and see for themselves if the German people were improving. Anyone should be able to see, he averred, that this "is the rebirth of a nation, brought about by the deliberate breeding of a new being [Mensch]."[88] As we shall see in the following two chapters, Hitler's fanatical pursuit of evolutionary progress and a "new man" would lead to increasingly radical solutions.

CHAPTER EIGHT

The Struggle for Living Space: War and Expansionism

Hitler's Early Views on Living Space

When Hitler touched off World War II in Europe by sending German forces into Poland on September 1, 1939, he was not just bent on regaining territory Germany had lost in World War I. His public proclamations that he needed to protect the German minority in Poland were a flimsy façade to justify his actions before a world that would never assent to his real plans of violent conquest, exploitation, deportation, and racial extermination. About three months before invading Poland, Hitler candidly told his highest military leaders in private that war against Poland was inevitable. However, contrary to his public statements, Hitler informed them, "It is not Danzig that is at stake. For us it is a matter of expanding our living space (*Lebensraum*) in the East and making food supplies secure and also solving the problem of the Baltic states. Food supplies can only be obtained from thinly populated areas."[1] Almost three months after the war began, Hitler told his military officials in a private speech that this war was a racial struggle caused by the growing German population. The goal was to bring the population size and the living space into harmony.[2]

Though one of Hitler's goals in gaining living space was economic—especially increasing food production—those who argue that Nazi expansionism was economically, but not racially, motivated misunderstand the whole thrust of Nazi *Lebensraum* ideology.[3] Yes, Nazi expansionism was intended to gain agricultural (and mineral) resources.

However, the primary goal was to gain territory for the expansion of the German race, as Hitler made clear repeatedly, especially in private speeches to military officers during the war. The economic goals were subservient to the racial ones, because economic plans always aimed at promoting the expansion of the German race. The intent was not to make the people already residing in Germany wealthier or increase their standard of living, though at times it might have done that. Rather, the point was to increase the German population by providing land for German settlers and food for a burgeoning German population.

Thus, ethnic cleansing or genocide was implicit in Hitler's vision of the struggle for living space, because the "inferior" races would have to be displaced to make room for the "superior" conquerors. Gerhard Weinberg is right to sum up Hitler's ideology thus: "The struggle for existence in which the races of the world engaged, the basic element of life on earth, was fundamentally a struggle for space. In this struggle the stronger won, took the space, proliferated on that space, and then fought for additional space. Racial vitality and spatial expansion were directly related."[4]

We have already seen that Hitler's belief that population expansion is necessary and beneficial was rooted in Darwinian ideology. We have also seen that the Darwinian struggle for existence—especially the struggle between races—played a central role in Hitler's worldview. In the Darwinian struggle for existence, organisms, especially those of the same species, compete for scarce resources to sustain an expanding population. Hitler followed the social Darwinist geographer Friedrich Ratzel in interpreting the struggle for existence as primarily a struggle for living space (*Lebensraum*), that is, a struggle for land needed to provide sustenance for a species or race. Since land could only be appropriated by conquest, Hitler believed that the struggle for existence among humans necessarily involved military conflict.[5]

Ratzel's leading disciple in the 1920s was Karl Haushofer, professor of geography at the University of Munich. Haushofer, a mentor and friend of Hitler's right-hand man, Rudolf Hess, admitted that Ratzel's ideas about living space were central to his own geopolitics. During the Third Reich, Haushofer published a selection of Ratzel's works, claiming that they were crucial in forming Nazi ideology. Without specifically mentioning Hitler, he also claimed that Ratzel's book, *Political Geography*, had been widely read by inmates in Landsberg in 1924.[6] Haushofer, who visited Hess seven times in Landsberg in 1924, specifically claimed that Hess had read Ratzel's book.[7] Thus, it is possible that Hitler imbibed his ideas about *Lebensraum* directly from Ratzel, but it is almost certain that he was influenced by Hess in this regard. Hess reported to a correspondent in July 1924 that the inmates were holding extensive discussions about *Lebensraum* at that time. This was precisely the time that Hitler was composing his chapter on "Munich," where he first clearly articulated his views on *Lebensraum*.[8] Though Hitler did not follow Haushofer in all the details of his geopolitics, he nonetheless did appropriate some central elements. Bruno Hipler, who overstates the importance of Haushofer in the development of Hitler's ideology, explains correctly that "the core of Haushofer's worldview was the social Darwinist *struggle for existence* between peoples (*Völker*) as the *ethically* valuable struggle for living space."[9] This was Hitler's view as well.

In his first several years as a politician, Hitler only rarely discussed Germany's need for more territory, and he did not yet publicly advocate military expansion toward the East to gain living space.[10] His foreign policy objectives seemed more like a combination of revanchism and Pan-Germanism. In his early speeches one of his favorite themes was that Germany needed to throw off the shackles of the hated Versailles Treaty. Most Germans were outraged by the treaty, which stripped Germany of territory and colonies. The first three points of the Nazis' Twenty-Five Point Program of February 1920 implied territorial expansion, but only to regain what Germany had lost in the war and to incorporate into Germany all territories with ethnic Germans (meaning primarily Austria and part of Czechoslovakia and Poland). By pressing for a Pan-German state the Nazis were going beyond the status quo ante bellum, but these were still limited goals. The third point stated, "We demand land and soil (colonies) for the nourishment of our people and for the settlement of our excess population." By mentioning colonies, this seemed to imply that the National Socialists were merely demanding the return of their former colonies in Africa and the Pacific.

However, by December 1922, at the latest, Hitler had embraced the view that overseas colonies were not what Germany needed. He told Eduard Scharrer that Germany needed to limit itself to a continental policy and not come into conflict with England. Rather, he stated, Germany should seek England's help in destroying Russia. This was because "Russia has sufficient land for German settlers and a wide field of activity for German industry." Thus, even before his incarceration at Landsberg, Hitler embraced German settler colonization in Eastern Europe, including Russia.[11]

It is unclear if Hitler favored this kind of expansionism before 1922, since he never clearly articulated it. In a December 1919 speech, however, he asked if it is right that a Russian has on the average eighteen times more land than a German. This rhetorical question implies that the present distribution of land is unjust. However, at this time Hitler did not dwell on this theme, nor did he overtly call for warfare to rectify this supposed injustice.[12] In November 1920 Hitler fulminated that after consolidating itself internally, Germany "can turn toward the East."[13] However, again it is not clear if he was only referring to the areas with German populations (thus equivalent with Pan-German goals) or to eastern territories further afield.[14]

By 1924 Hitler publicly committed himself to an ambitious program of militarism and expansionism. During his trial for treason in 1924 he told the court why he thought the Weimar Republic's peaceful foreign policy was misguided and debilitating: "The preservation of world peace can never be the purpose and means of the political leadership of a nation, but rather eternally the only goal and purpose can be the multiplication and preservation of a *Volk*." He then claimed that because its aim is to expand the population, the "goal and purpose of the state is not limited." In the same speech Hitler also vented his spleen at the French, claiming that in World War I the French had waged a "war of destruction, a racial struggle," whose goal was "to eliminate twenty million Germans from Europe." By this time, if not earlier, Hitler clearly believed that population expansion necessarily produced military conflict between races and states.[15]

In *Mein Kampf* Hitler expostulated in detail about why he thought expansionism was the only prudent policy for Germany. In the opening passage of his book he promoted his Pan-German goal of uniting Germany and his homeland of Austria. All ethnic Germans must belong to the same state, he declared.

> Never will the German nation possess the moral right to engage in colonial politics until, at least, it embraces its own sons within a single state. Only when the Reich borders include the very last German, but can no longer guarantee his daily bread, will the moral right to acquire foreign soil arise from the distress of our own people.[16]

Here Hitler clearly gave his Pan-German goals priority over other territorial expansion. However, he ominously implied that further expansion would be morally justified after his Pan-German goals were met.

A little later in *Mein Kampf* he explained in great detail why Germany needed to pursue an expansionist foreign policy. He explained that foreign policy must take into account the present population growth of 900,000 people per year. He then discussed four possible solutions: (1) birth control to reduce population growth; (2) inner colonization; (3) acquisition of new land; or (4) expansion of industry in order to import more food. We have already seen that Hitler rejected birth control, because he thought it violated the laws of nature, particularly the Darwinian law of natural selection (see chapter 6). He rejected inner colonization, because he thought that limiting one's territory would only give an advantage to less cultured races in the struggle for existence. He believed that inferior races were more brutal and, if possible, would expand their territory at the expense of the more cultured races. Hitler also rejected industrial expansion as a permanent solution, since it too would result in inevitable conflict, especially with Britain. Hitler was always skeptical about the long-term reliability of imports, since the British had blockaded Germany during World War I.[17]

For Hitler the only viable solution was to seize more land, so Germany could grow its own foodstuffs and not have to rely on imports. He did not believe that present borders were just and needed to be maintained. Rather,

> Nature as such has not reserved this soil for the future possession of any particular nation or race; on the contrary, this soil exists for the people which possesses the force to take it and the industry to cultivate it. Nature knows no political boundaries. First, she puts living creatures on this globe and watches the free play of forces. She then confers the master's right on her favorite child, the strongest in courage and industry.

According to Hitler, the land Germany needed was in Europe, not in Africa or overseas. Specifically, he wanted Germany to expand at Russia's expense, since that is where land was more sparsely populated.[18]

Hitler expanded on these ideas in the penultimate chapter of *Mein Kampf* on "Eastern Orientation or Eastern Policy." He explained once again that German foreign policy must aim at the preservation of the Aryan race by gaining the requisite territory to support an expanding population. The "highest aim of foreign policy" is "*to bring the soil into harmony with the population.*" After explaining this point, Hitler then declared: "I still wish briefly to take a position on the question as to what extent the demand for soil and territory seems ethically and

morally justified." After circuitously attacking the inadequacy of pressing for the borders of 1914, he explained the moral underpinnings of
his expansionist program:

> For no people on this earth possesses so much as a square yard of
> territory on the strength of a higher will or superior right. Just
> as Germany's frontiers are fortuitous frontiers, momentary fron-
> tiers in the current political struggle of any period, so are the
> boundaries of other nations' living space. And just as the shape
> of our earth's surface can seem immutable as granite only to the
> thoughtless soft-head, but in reality only represents at each period
> an apparent pause in a continuous development, created by the
> mighty forces of Nature in a process of continuous growth, only
> to be transformed or destroyed tomorrow by greater forces, like-
> wise the boundaries of living spaces in the life of nations. *State
> boundaries are made by man and changed by man.*[19]

Thus Hitler morally justified territorial expansion by appealing to the
evolutionary laws of nature. He equated strength and success in the
struggle for existence with moral justice.

As indicated by the title of his chapter, Hitler believed that expansion toward the East was the fate of Germany. The Bolshevik seizure of
power was a serendipitous development that would ultimately benefit
Germany, he claimed, since the Bolsheviks destroyed the best racial
elements of the Russian people. According to Hitler the Bolsheviks had
put into power Jews, who were unable to organize a state effectively.
As a result, Russia was much weaker. Furthermore, Hitler claimed that
Jewish-Bolshevik domination of Russia made war with Russia inevitable, since this was a step on the path to Jewish world domination.
Their next goal, he maintained, would be to destroy Germany.[20] Thus
Hitler's goal of territorial aggrandizement at the expense of Russia
would not only benefit Germany racially by providing more living
space for population expansion, but it would also destroy their racial
archenemy: the Jews.

After being released from Landsberg Prison in 1924, Hitler continued to speak and write about the necessity of gaining living space.
In a flyer published in December 1925 on "The Social Mission of
National Socialism," he made clear that in the long run an expansionist foreign policy was his solution for Germany's economic travails. He
ascribed Germany's present economic woes to overpopulation and lack
of land. He complained that the problem of overpopulation had led

to emigration in the past, which had robbed Germany of some of its most valuable blood. He then summed up his foreign policy goals as: "*Preservation and progress, sustenance and protection of our Volk and the most valuable racial elements in this Volk. That is the exclusive and only goal!*" He then maintained that the only ideals he could support were ones that brought victory to the Aryan racial elements in the struggle between peoples.[21]

Hitler broached the topic of the need for more living space in many speeches in the mid- to late 1920s. He insisted that land and bread could only be won by military means.[22] In 1929 he told the Nuremberg Party Congress that the National Socialist emphasis on the problem of space had forced other parties to begin discussing the problem. However, their solutions to the problem were inadequate, because they saw it only as an economic issue. "When I as a National Socialist take up the subject of this vital question for the German *Volk*," he asserted, "then I treat it not for economic purposes, but for purposes of race hygiene, for the purpose of preserving the future power of our *Volk*, indeed for preserving the *Volk* itself." Gaining living space, then, was always linked to the preservation and improvement of the race.[23]

The struggle for living space played a central role in Hitler's unpublished *Second Book*. At the very beginning of the book he explained that the reproductive drive is unlimited, but space is limited. "In the limitation of this living space lies the compulsion for the struggle for survival, and the struggle for survival, in turn, contains the precondition for evolution," he stated. A few pages later he claimed that "a people's [*Volk's*] entire struggle for survival in reality consists only of securing the necessary territory and land as a general precondition for feeding the growing population." While admitting that war was a necessary means to obtain this living space, Hitler denied that war was a purpose of life. It was a necessary means, not a goal. He even warned about pursuing a policy of perpetual war, which would be just as catastrophic as a policy of perpetual peace.[24] Thus, Richard Bessel's claim in his excellent recent book, *Nazism and War*, that war was "the essence of the Nazi project" is not quite right.[25] Bessel correctly demonstrates that war was an integral part of the Nazi project, but for Hitler war was not an end in itself. Rather, in Hitler's worldview war was supposed to bring about evolutionary advance by providing living space for the highest race on the globe. It would elevate the human species and also thereby lead to cultural progress as well.

Just as in *Mein Kampf*, Hitler once again addressed the moral question in his *Second Book*: Is it not immoral to take land away from others?

Again he provided essentially the same answer: "Therefore, every healthy native people sees nothing sinful in the acquisition of land, but rather something natural. The modern pacifist, however, who repudiates this most holy right" lives off past injustices. Borders are constantly changing, just as organisms are evolving, he stated. The earth has not been given to anyone in perpetuity, he declared, but belongs to whomever has the strength and courage to seize it. The moral question can only be decided by the struggle for existence, because "the first right in this world is the right to life, provided one has the strength for it." By the "right to life" Hitler did not mean that everyone had some intrinsic right that others should not violate. On the contrary, he meant that each individual and race had the right to preserve its own life, even if this entails stamping out the life of others. He again indicated that Germany would have to seek its living space in the East.[26]

Expansionist Agenda during Peacetime

When Hitler came to power in 1933, he shrewdly proclaimed himself a man of peace and publicly eschewed expansionist goals. He recognized the need to placate foreign diplomats and political leaders, lest they stymie his radical foreign policy objectives. However, Hitler never abandoned his expansionist ideology. He was merely biding his time. Just four days after being appointed chancellor, Hitler met with Germany's military leadership. He was very candid about his expansionist agenda, informing them that the way to solve the present unemployment problem was through a settlement policy that "presupposes an expansion of the living space of the German *Volk*." He predicted that the internal struggle against Marxism would take six to eight years, after which Germany could pursue an active foreign policy. Thereby, he explained,

> the goal of the expansion of the living space of the German Volk would be achieved by force of arms—the goal would likely be the East. However, the Germanization of the population of annexed or conquered land is not possible. One can only Germanize the land. One must ruthlessly deport a few million people like Poland and France did after the war.[27]

Thus Germany's military leadership knew from the first days of the Nazi regime that Hitler was bent on waging a war of aggression in the

East that would involve deporting foreign populations to make room for German settlers.

Between 1933 and 1937 Hitler never publicly advocated war or expansionism. He rarely (if ever) even used the term "living space" (*Lebensraum*) in public. In his first year in power he was asked point blank by a British journalist in an interview about the phrase, "*Volk without Space,*" which was causing angst in British circles. Hitler conceded that Germany was overpopulated, and he suggested that other powers should make concessions to them because of it. However, he vigorously denied that Germany would resort to arms, insisting instead that Germany would rely only on peaceful negotiations.[28]

In May 1937—with Germany already well under way in its rearmament program—Hitler began speaking publicly once again about the need for living space. During his May Day speech that year he praised German laborers for their diligence and ability. However, the German *Volk* "is living in a space much too tight and too confined to possibly provide it everything it needs," he continued. Hitler then maintained that because of its lack of space the "struggle for life" is more difficult for Germans than for other peoples. He continued, "Life itself puts every generation under an obligation to wage its own battle (*Kampf*) for that life." Though Hitler did not specifically mention war as a means for gaining living space, by invoking the "struggle for life" he was moving ever closer to divulging his real aims: offensive warfare.[29] In other public speeches in 1937 and thereafter he also stressed the need for more living space, though before the outbreak of World War II he never openly indicated war as the necessary means.[30] In his May Day speech in 1939, for instance, after stressing the importance of living space, he asserted, "The highest command for us is the securing of German Lebensraum." Then to allay fears that this might arouse, he immediately (and hypocritically) proclaimed his commitment to peace. Even taking Hitler's hyperbole into account, it is evident that acquiring living space was a moral imperative for Hitler.[31]

Hitler's private speech on November 23, 1937, to the Adolf Hitler School in Sonthofen, where the Nazis were prepping future German leaders, was almost exclusively about Germany's need for living space. Hitler rattled off statistics about the amount of land controlled by various European countries, the United States, China, Brazil, and Japan. Then he compared the populations of other European countries with Germany's population. The lesson was elementary: Germany was getting the short end of the stick and needed to become more assertive in foreign policy. Otherwise, he warned, "*Our lack of space will result in the death of our Volk.*" Germany has a moral right to pursue more living space, he asserted, but ultimately only power can decide who has the moral right to land. *Though Hitler stopped short of openly advocating war, most of his audience probably comprehended the thinly veiled point.*[32] Again we see that Hitler regarded expansionism as a righteous cause.

By 1937 Hitler was already secretly preparing for war. The primary goal of the vaunted Four Year Plan was to prepare the German economy for war, though the Nazis never admitted this publicly. Hitler, however, divulged the real purposes of the Four Year Plan in a secret essay he wrote in 1936, the year the plan began. In the first sentence of the essay he explained that his views revolved around the "struggle for life between the peoples (*Völker*)." Then, after discussing the threat of the Soviet Union and Bolshevism, Hitler called for a German military buildup. The economy must be subservient to the *Volk*, providing for its preservation and protection. Up to this point in the essay Hitler's concerns seemed defensive, but toward the end, Hitler laid bare his expansionist agenda. As in his earlier speeches and writings, he claimed Germany was overpopulated and the solution to overpopulation was the "expansion of living space." He closed the essay by demanding that the German army and economy be ready for war in four years.[33]

Under the Four Year Plan the German government began investing massive sums to beef up production of steel, heavy machinery, and other goods needed to increase armaments production. When Hitler's Economics Minister, Hjalmar Schacht, warned him that military expenditures needed to be scaled back, he sacked him and continued to spend lavishly on the military. Richard Overy argues that by 1938–1939 the massive scale of economic mobilization for arming Germany indicates that Hitler intended more than just localized Blitzkrieg wars with Czechoslovakia and Poland. His long-term strategy included a major war with the leading powers of Europe to gain living space. Overy states, "Economic questions, when considered at all, were subsumed into his [Hitler's] great plans for the future; the plans for *Lebensraum* and the plan to wage a 'life-and-death struggle' for the survival of the race."[34] Overy's position is confirmed by Hitler's statement to his military leaders in May 1939 that they should do everything possible to ensure a brief war, but yet be prepared for a war lasting ten to fifteen years.[35]

By 1937–1938 the German economy faced the enviable position of needing more labor to sustain its growth, while most other industrialized countries were still in the grips of depression. In order to get the German economy ready for war, Hitler willingly set aside his earlier policy of getting women out of the workforce. In 1938, for instance, the Nazi regime altered the marriage loan policy, henceforth allowing couples to get marriage loans even if the woman continued working outside the home. Nazi propaganda also began encouraging women to work outside the home, especially once World War II was underway. Hitler was not primarily concerned about women's traditional roles. He did not care whether women stayed at home or worked in a factory or office. His sole aim was more German babies and more territory to support an expanding population. If women working in industry could help Germany win the struggle against surrounding races and gain more living space, then he was all for it. However, later during the war Hitler refused to conscript women to work for the war effort, because he thought this would harm birthrates.[36] Here, then, is another example of an inconsistency in Nazi policy that was nevertheless ideologically driven. Higher ideological goals always trumped transitory policy decisions, sometimes leading to inconsistencies and wavering in specific Nazi policies.

Planning and Waging a War for Living Space

On November 5, 1937, Hitler met with his Foreign Minister, War Minister, and the commanders-in-chief of the army, navy, and air force to discuss future military plans. He opened the meeting by telling them that the "aim of German policy was to make secure and to preserve the racial community and to enlarge it. It was therefore a question of space." The size of the German population "implied the right to a greater living space." This living space must be acquired in Europe, not in overseas colonies, and it could only be achieved by military force. At this meeting Hitler mentioned only the conquest of Austria and Czechoslovakia; he did not yet divulge his plans for Poland or the Soviet Union. Nonetheless, he did explain how the conquest of Austria and Czechoslovakia would benefit Germany. He estimated that Germany would gain enough land to provide sustenance for five or six million more people, "on the assumption that the compulsory emigration of two million people from Czechoslovakia and one million people from Austria was practicable."[37]

Before this time Hitler had usually avoided discussing the obvious implications of his call for more living space. In *Mein Kampf* he proposed that when Germany acquired new territories they should only be populated by settlers who had certificates vouching for their racial purity. Establishing these "border colonies" with racially superior Germans would produce the "germ for a final, great future evolution of our own people, nay—of humanity."[38] What Hitler conveniently omitted from this scenario—and what he neglected to admit on most occasions when he spoke about the need for living space—was that forcibly expelling native populations from those "border colonies" was the only way to make this evolution possible. His vision of gaining living space always implied that the native populations would have to be eliminated to make room for German settlers. He had stated this plainly in his *Second Book*, too, though it was not published, so his contemporaries would not have known it. In discussing what to do with the "alien racial elements" in Poland, he proposed that either they would have to be isolated to prevent racial mixing, or else they would have to be deported so the land could be redistributed to ethnic Germans.[39]

When Germany forced Czechoslovakia to surrender without a fight in March 1939, Hitler declared that the Czech provinces of "Bohemia and Moravia have belonged to the Lebensraum of the German *Volk*." He fully intended to Germanize these lands by deporting those deemed racially inferior, while rechristening any Czechs with suitable racial traits as Aryans. The Agricultural Minister Walther Darré had proposed to Hitler already in 1930 that Germans should settle Slavic lands in the East, a plan that corresponded perfectly with Hitler's own ideas.[40] In February 1937 Hitler had personally commissioned an official in the Agricultural Ministry to draw up secret plans to resettle German farmers in Czechoslovakia (and the Ukraine).[41] In October 1940 the Nazi leaders in Bohemia and Moravia presented a plan to Hitler to Germanize their territories. Hitler approved this plan, which called for expelling about half the Czechs, while allowing the other half to join the German *Volk*. The mass deportations of Czechs never transpired, however (except for the Jews), since there simply were not sufficient German settlers available and Germany needed their labor.[42] Also, by the fall of 1939 Germany's resettlement efforts concentrated more on Poland than on Czechoslovakia.

Ten days before touching off World War II by invading Poland, Hitler instructed his generals that this campaign would be different from old-fashioned European warfare. One of the generals present recorded in his diary that Hitler told them to destroy the enemy "harshly and ruthlessly! Steel yourselves against all considerations of sympathy!" Hitler exhorted his commanders to wage war brutally, instructing them that the goal of the war against Poland was not reaching a certain line, but rather destroying and eliminating the enemy. The means to achieve this are irrelevant, for "victors are never questioned whether their reasons were justified. It is not a matter of having the right on our side, but solely of gaining victory."[43]

Wartime brutalities in Poland were not just the side effect of a harsh military campaign that Hitler orchestrated. Rather, Hitler planned atrocities before the military campaign began. He ordered Reinhard Heydrich to organize SS and police units to move into Poland behind the German army to arrest and execute various racial and political opponents, including Jews, Freemasons, Catholic clergy, and socialists. They carried with them lists with 61,000 names of people slated for death. By December 1939 the German police units had executed about 50,000 Poles, of whom 7000 were Jews. Death to the native populations was clearly part of Hitler's war planning.[44]

As the German military finished up its ruthless campaign against Poland, Hitler hatched plans for the racial reordering of Poland. In late September Hitler expressed the desire to carve Poland up into three zones: (1) the former German territory, which would be fully Germanized by settling ethnic Germans there; (2) a central zone extending east to the Vistula River, which would contain the "good Polish elements"; and (3) a zone east of the Vistula for inferior Poles and Jews.[45] The Nazis did not strictly follow this tripartite plan, though they did try to set up the first zone by annexing a large chunk of western Poland. In four separate deportation actions from late 1939 to early 1941, about 300,000 Poles were deported from a portion of the newly annexed territory known as the Wartheland so that ethnic Germans from other Eastern European countries could take their land.[46] Hitler did not consider the borders drawn up in this plan for Poland permanent, since he also said when he laid out his plan, "The future would show whether after a few decades the cordon of settlement would have to be pushed farther forward."[47] Hitler's quest for living space was limited only by the ability of Germans to reproduce enough to populate those territories.

On October 7, 1939, Hitler appointed Himmler to the position of Reich Commissar for the Strengthening of the German *Volk*. This gave the SS—especially its Race and Settlement Office—a leading role in organizing the deportation and resettlement schemes in Poland. It also signaled a policy decision by Hitler to accelerate the deportation and resettlement of Poland rather than following Darré's plans for a more gradual resettlement. Himmler had police forces at his disposal, too, to execute the deportations.[48]

Hitler stayed well-informed about the resettlement activities in Poland. In March 1940 Arthur Greiser, the Nazi leader of the Wartheland, reported to Hitler that he had succeeded in increasing the German population in the city of Posen from 2000 to 50,000.[49] Hans Frank, Nazi ruler of the Generalgouvernement, which comprised German-occupied Polish territory east of the annexed territory, constantly complained to Hitler about the stream of deportees being shunted his way from the annexed parts of Poland. Hitler had initially told Frank "to plunder his lands mercilessly," but by early 1940 he shifted to exploitation of Polish labor to benefit the German war effort.[50] This shift to using the Poles as slave labor was not an opportunistic abandonment of ideological goals, as some historians have claimed, for even before the Polish campaign Hitler had informed his military leaders that the Polish population would be available for "labor service."[51] Rather it signaled that leading Nazis—including Hitler—did not always know how to achieve their ideological goals, which sometimes led to chaotic plans and policies.

Hitler was not particularly responsive to Frank's pleas to end the deportations to the Generalgouvernement. In November 1940 he told associates that Frank would simply have to accept all the "riff-raff" being sent his way, because his territory was needed as a labor reservoir for the time being.[52] In March 1941, however, Frank reported to his subordinates that Hitler had promised him that the Generalgouvernement would ultimately be Germanized. Hitler pledged to replace the twelve million Poles with four or five million Germans in the next twenty years.[53] Frank was elated. The Poles, however, were doomed if these plans had come to fruition. Hitler and other Nazi planners probably did not have in mind systematic extermination for the Poles.[54] However, even mass deportations had genocidal overtones, since shunting large populations further east would likely result in mass death.

Hitler's war for living space was thrown offtrack by a fatal miscalculation. Hitler did not think the British and French had the backbone to fight over Poland. Thus, when Britain and France declared war on Germany on September 3, 1939, Hitler was forced into a war he did not want. The war in the West was more conventional, since Hitler was not bent on expelling residents to clear space for German settlers. However, the racial struggle reared its ugly head even in the Western campaign. While treating white French troops according to the Geneva Convention, German troops massacred thousands of black African colonial troops fighting for France, including defenseless POWs. Some of these massacres took place under orders from German officers. There is no evidence that Hitler or other high-ranking Nazis ever issued orders to slaughter blacks in France. However, Goebbels—with Hitler's approval—did initiate a propaganda offensive against black colonial troops during the French campaign that dehumanized blacks and seemed to give official sanction to commit atrocities against them.[55] The official SS newspaper argued in an article during the French campaign that blacks "stand on a level of evolution not much higher than the gorilla."[56] No German soldier was ever punished by their superiors for massacring blacks.

Wresting Living Space from the Soviet Union

After waging successful campaigns against Denmark, Norway, Holland, Belgium, and France in the spring of 1940, and after failing to bring Britain to her knees, Hitler prepared for war against the Soviet Union. Though some historians point to tactical and pragmatic considerations behind Hitler's decision to attack, most concede that Hitler's *Lebensraum* philosophy played a crucial role in the decision-making process. In his private monologs Hitler clearly interpreted the Eastern campaign as a struggle for living space. In September 1941, as German armies advanced through Soviet territory, Hitler told his colleagues that the real dividing line between Europe and Asia would not be the Ural Mountains, but would rather be the border between the Germanic and the Slavic peoples, which the Germans would determine. He argued that it was unreasonable for the superior Germans to have so little space, while the inferior Russians, who have no use for culture, have huge expanses. He continued, "We must create conditions that allow our *Volk* to reproduce, but that restrict the reproduction of Russians." The war against the Soviet Union was necessary, he asseverated, to enable the German population to continue increasing.[57] A couple of weeks later Hitler argued that the present campaign against the Soviet Union represented a war that had returned to its most primitive form: a war for space. Originally, he asserted, wars were fought over access to food. These wars "correspond to the principle of nature, ever anew to bring about selection through struggle: The law of existence demands uninterrupted killing, so that the better one lives."[58]

Indeed Hitler fully intended that this campaign should depopulate Soviet territory. Three months before the invasion, Hitler told his leading generals that the war in the Soviet Union would not be like the war in the West. Rather it would be a harsh war of annihilation (*Vernichtungskampf*).[59] Hitler had apparently already conveyed that message to other generals, for a few days earlier General von Brauchitsch had informed other senior military leaders that troops "have to realize that this struggle is being waged by one race against another, and proceed with the necessary harshness."[60] In May 1941 the Nazi regime and the military agreed to a "hunger plan" for the Soviet Union, whereby grain would be requisitioned to feed Germans, making starvation in Soviet territories inevitable.[61] Millions of Soviet POWs died of starvation or were shot by German firing squads. Hitler planned to utterly decimate Soviet cities. When the Germans attacked Kiev in August 1941, Hitler ordered that the air force reduce it to rubble. Apparently the air force did not have sufficient bombs to comply, so the Nazis began reducing the population of Kiev by a conscious starvation policy.[62] Hitler told Goebbels on several occasions that he favored starving out Moscow and Leningrad, making further life in these cities "a misery and a hell." After eliminating the populations, he planned to completely wipe out the cities, returning them to the plow.[63] Nazi brutality was not just aimed at crushing the Soviet Union militarily, but also brutally depopulating its territory.

Hitler hoped the depopulation of occupied territories in the Soviet Union would lead to German colonization. He gloated to Goebbels in November 1941 that Germany would settle the Crimea with the "best German human material" and annex it to Germany. A couple of weeks later he informed high-ranking Nazi leaders (*Gauleiter*) that the lands in the East would be Germany's India. "That is our colonial land that we want to settle," he bragged. He again mentioned the Crimea as prime territory for Nordic settlement. In three or four generations, he optimistically prophesied, the lands in the East will be completely German.[64] In May 1942 he told his entourage that the long-term goal of his Eastern policy was to settle about 100 million Germans there.[65] Two weeks later he told Nazi officials if Germany followed a wise population policy that included the reintegration of Germans who had earlier emigrated (to the United States, for instance), within seventy or eighty years Germany should have a total population of about 250 million.[66] Obviously, these were grandiose plans and Hitler knew they would take decades or more to fully implement.[67] Nonetheless, even though Germany only controlled this territory for about three years, Himmler and his subordinates began formulating plans to Germanize Soviet territory. In the summer of 1943 Himmler drew up plans to resettle the ethnic Germans already living in the Ukraine. He wanted to concentrate them in strategic locations that would serve as nuclei for future German settlement.[68]

Promoting His Lebensraum Philosophy during the War

During World War II Hitler often referred to the war as a struggle for existence and mentioned the need for living space, both in public and in private. However, only in his private speeches did he explain in great detail his philosophy of war. Between November 1939 and June 1944 he delivered at least seven major private speeches to military officers and officer cadets. In all of these, Hitler's *Lebensraum* philosophy occupied a central position. Since the ideas were so similar in these speeches, I will only analyze two of them. However, all seven speeches contain many of the same ideas, focusing on the struggle for existence, population expansion, and the need for living space.[69]

In his private speech to military officers in June 1944, Hitler candidly divulged his philosophy of war. The opening words of his speech reveal his Darwinian mindset:

> War belongs to those events that are essentially unalterable, that remain the same throughout all times and only change in their form and means. Nature teaches us with every insight into its functioning and its occurences that the principle of selection rules over it, that the stronger remains victor and the weaker succumbs. It teaches us that what often appears to an individual as brutality, because he himself is affected or because through his education he has turned away from the laws of nature, is nonetheless fundamentally necessary, in order to bring about a higher evolution of living organisms.

Humanitarianism and any kind of weakness will lead only to doom and could lead to the extinction of the entire human species. Because of this, war is not only necessary, but good. "War is thus the unalterable law of all life, the precondition for the natural selection of the stronger and at the same time the elimination of the weaker. What appears to people as brutal is from the standpoint of nature obviously wise." Hitler intended by this war to eliminate those he considered weaker and inferior races to pave the way for higher evolution for the human species.[70]

In a private speech in February 1942 to military officers, Hitler clearly explained the link between his Darwinian vision and living space:

> Nature gives to all organisms the urge to reproduce; self-preservation and reproduction are the two most natural instincts, which all

living organisms possess. And nature places no limits on either of them. The limitation occurs only through the struggle for life, that is, through the power of the organisms themselves. For people this means the following: Normally a *Volk* will grow without ceasing. If the natural laws of its reproduction suffices, a people must necessarily increase numerically; if on the other hand, its living space remains the same, then in the course of time again and again a misproportion between the increasing population and the constant living space will emerge.[71]

If this misproportion is not rectified, he stated, it can lead races to extinction. Thus humans must struggle to gain new living space.

Since this new living space can only come at the expense of people already occupying it, this Darwinian struggle necessarily involves the death of weaker people, who have to vanish to make way for the stronger ones: "In that one individual lives, he hinders the life of another; and in that he dies, he makes the path clear for the life of a new individual."[72] Hitler saw life as one gigantic struggle-to-the-death between individuals and especially between races. War was a necessary part of this natural process. Even though it may seem unpleasant or immoral, war nonetheless served a beneficent purpose by ridding the world of inferior people and thus improving the species.

Hitler's speeches, together with other Nazi propaganda, were pretty successful in inculcating these ideas into the German army. Geoffrey Megargee explains,

That the quest for Lebensraum would require aggressive war was a fact that the military accepted as a matter of course, for a variety of reasons. First, a growing number of right-wing Germans believed in a loose ideology that we now know as Social Darwinism, which applies elements of Charles Darwin's evolutionary theory of biology to human societies. According to this belief system, conflict is inevitable, between individuals and nations as between animal species, and only the strong deserve to survive. This was a fundamental principle of Nazism.[73]

Omer Bartov has also shown that the German army imbibed a great deal of Nazi ideology and internalized many elements of its racial thought.[74]

Despite Hitler's continual glorification of war and his megalomaniacal pursuit of territorial aggrandizement, he did recognize that war could have negative consequences. Many eugenicists in the early twentieth century warned that war could lead to the death of the brightest and the best, causing biological degeneration. Hitler recognized this problem already in *Mein Kampf*, where he claimed that the revolution of 1918 had occurred because the best young Germans had died at war or were still at the front, while the inferior elements stayed at home and led the revolution.[75] In his *Second Book* he explained that the dysgenic effects of war showed the foolishness of a policy of constant war. In war the "hero dies, the criminal survives," he asserted. A statesman's enthusiasm for war must thus be tempered, since too many wars could "gradually bleed a people of its best, most valuable elements." He then warned that wars that cannot ensure a replacement of the dying troops are a sin against the *Volk*.[76] In the mid-1930s Hitler assured people that the dysgenic effects of war made him committed to peace.[77] During World War II Hitler continued to express concern about the "negative selection" caused by war.[78]

In the early phases of the war—at least until mid-1942 and perhaps much longer—Hitler had overweening confidence that the losses Germany incurred during World War II would be offset by the population increases made possible by the new living space at their disposal. In September 1941 he told Goebbels that despite the losses on the Eastern front, Germany would be able to make up the losses by an "enormous population increase." He unrealistically estimated that Germany would lose only about 10 percent as many men as it did in World War I.[79] In January 1942 he myopically calculated that if the war cost Germany a quarter of a million dead, it could make that up by excess births. "Our salvation will be the child!" he exclaimed. Our losses "will be resurrected in multiplied numbers in the settlements, which I am creating for German blood in the East."[80] In May 1942 Hitler continued to justify the casualty figures by arguing that Germany had lost far more people by restricting births in the post-World War I period than he had sacrificed in war.[81] Hitler never showed much concern about the deaths he was causing by war anyway. In August 1941, in the midst of a discussion about war casualties with some colleagues, he stated, "Life is brutal. Coming into existence, existing, and passing away, there is always killing; everything that is born must die again, whether through illness, accident or war, it is all the same." Those dying in war, however, should find comfort in knowing that their sacrifice will benefit their *Volk*, he reassured his entourage.[82]

Hitler also had another solution to the problem of "negative selection" during war: murder. If multitudes of manly young men, the cream of the crop in Germany, were perishing at the front, he hoped to offset their deaths by killing the allegedly inferior people who could not fight. Killing the disabled and racial inferiors was linked to his program of military expansion to ensure the biological vitality and improvement of the race and ultimately all of humanity, as we shall see in the following chapter.

CHAPTER NINE

Justifying Murder and Genocide

Murdering the Disabled

Only once in his life did Hitler ever sign a document authorizing murder. Usually when he wanted to bypass legal niceties to crush his political opponents or destroy his racial enemies, he merely issued verbal directions to his loyal minions. In October 1939, however, he signed a brief memo to Philipp Bouhler, head of the Chancellery of the Führer, and to his personal physician, Karl Brandt. The entire memo stated:

> Reichsleiter Bouhler and Dr. Brandt are commissioned to extend the authority of those physicians they designate, so that mercy killing may be administered to those who according to human judgment are incurably sick, after diagnosis of the condition of their illness. [signed] A. Hitler[1]

Hitler probably signed this document to reassure physicians and other participants in this program that they would not face prosecution, since mercy killing was illegal in Germany. Hitler knew that killing the disabled was too controversial to legislate at that time, so he opted for secrecy. The document was so top secret that Hitler's Minister of Justice did not receive a copy until over ten months later, after he complained that no legislation had been passed permitting it. By that time many thousands of institutionalized mentally and physically disabled patients had been murdered in this so-called euthanasia program, which was code-named the T-4 Program (after its headquarters in Tiergartenstrasse 4 in Berlin).

Run directly under Hitler's auspices by the Chancellery of the Führer, the T-4 program set up six centers throughout Germany for murdering the disabled. Bouhler apparently kept Hitler apprised of the program, for Goebbels recorded in his diary that Bouhler informed Hitler about the progress of the "liquidation procedure for the insane" in May 1940.[2] By August 1941, when Hitler ordered the centers to close, about 70,000 patients had been killed by carbon monoxide gas, injections, starvation, or other means. After August 1941, until Germany's defeat in May 1945, the physician-directed murders continued, but in a more decentralized manner. By 1945 a total of about 200,000 disabled patients had been murdered by German medical personnel. In a few cases physicians continued murdering the disabled even after Germany had surrendered.

Historians who have analyzed the T-4 program recognize that it was a continuation and radicalization of previous eugenics policies.[3] The historian Hans-Walter Schmuhl not only highlights the tight links between Darwinism, eugenics, and euthanasia ideology, but he also explains, "The racial-hygiene paradigm constituted an ethic of a new type, which was ostensibly grounded scientifically in Darwinian biology." Besides replacing Judeo-Christian ethics with an evolutionary ethic, Darwinism influenced euthanasia ideology in yet another way, according to Schmuhl: "By giving up the conception of humans as the image of God through the Darwinian theory, human life was construed as a piece of property, that—contrary to the idea of a natural right to life—could be weighed against other pieces of property." Schmuhl argues that under the influence of Darwinism, a shift in thinking about the value of human life gave impetus to euthanasia.[4] Historians who have analyzed the euthanasia movements in the United States and Britain likewise stress the importance of Darwinism and eugenics in influencing the early euthanasia movement.[5]

The first German scholar to seriously propose killing the disabled was Haeckel. In the second edition of *The Natural History of Creation* (1870), his extremely popular exposition of Darwinian theory, he expressed support for eugenics. He lamented that some aspects of modern society were leading to biological degeneration. Then he insinuated that following the ancient Spartan practice of infanticide for weak and sickly babies might be beneficial.[6] In a later book in 1904 he explicitly sanctioned infanticide for the congenitally disabled, as well as involuntary euthanasia for the mentally ill. Physicians should decide whether such people's lives should be ended, he thought.[7]

Haeckel was not the only one to claim Darwinian sanction for killing the disabled. The prominent psychiatrist August Forel opened his book

The Sexual Question (1905) with a long discussion of the relevance of Darwinism for sexuality. Forel believed that Darwinism underpinned not only eugenics but infanticide for the disabled. He even referred to the mentally disabled as "little apes" in this work, borrowing a trope from Karl Vogt.[8] The Darwinian biologist and anthropologist Vogt had denigrated the mentally disabled as atavistic creatures who had not developed into fully human form.[9] Other eugenicists thinking that infanticide or involuntary euthanasia were beneficial forms of eugenics included the physician Agnes Bluhm and the medical professor Alfred Hegar.[10]

Debate over euthanasia became inflamed in the 1920s in response to the provocative book, *Permitting the Destruction of Life Unworthy of Life* (1920), coauthored by the legal scholar Karl Binding and the professor of psychiatry Alfred Hoche. Though most physicians in the 1920s continued to resist euthanasia, a significant minority of physicians, medical professors, and scientists embraced the idea. Many of them justified it on the basis of their understanding of Darwinism and/or eugenics. We do not know if Hitler read any of the above-named proponents of infanticide and euthanasia. However, by the 1920s the idea was circulating widely, and Hitler could have imbibed the idea from any number of sources.

Hitler agreed with those physicians and public health officials who characterized the disabled individual as a "life unworthy of life." He considered them inferior beings that before the advent of Christian and humanitarian ethics would have died out in the struggle for existence. Their deaths, Hitler believed, would benefit humanity, so he saw no reason to balk at giving them a speedy death. Thus, the goal of the T-4 program was the same as the goal of compulsory sterilization: to rid the German people of the unhealthy elements, making the Aryan race stronger and healthier.

The involuntary euthanasia program was connected in some ways to the war effort. The first mass killing of the disabled took place in November 1939 in occupied Polish territory that Germany annexed. Inmates of an asylum were gassed with carbon monoxide. Hitler also linked the T-4 program directly to the war effort by dating the secret memo to Bouhler and Brandt to the first day of the war, September 1, 1939, over a month earlier than he actually signed it. Hitler intended the T-4 Program to free up resources, especially money and hospital space, for the German war effort. A statistician in the Interior Ministry wrote a report in 1941 calculating that by killing about 70,000 disabled people up to that time, the T-4 program would ultimately save over

885 million marks over a ten-year period.[11] Nazi propaganda relating to the sterilization program often stressed the fiscal savings involved in reducing the number of disabled people in Germany. However, though economic arguments may have been one incentive for killing the disabled, Hitler ignored the economic argument whenever he discussed (usually privately) the reasons for euthanasia. For Hitler the economic goals were always subsidiary to the main purpose of the program: biological improvement and advancing the evolutionary process. Even if economic arguments had significantly influenced Hitler's decision to proceed with the T-4 program, Hitler's economic agenda was shaped by evolutionary considerations. Winning the racial struggle for existence was paramount, and economics was a means to that end.[12]

For Hitler, killing the disabled was supposed to help correct a biological problem the war created. He often remarked during the war that Germany's strongest and healthiest men were dying at the front, while the weak and sick stayed behind. In August 1942 he stated, "Every war leads to a negative selection. The positive ones die en masse."[13] He wanted to offset this negative selection that seemed to fly in the face of natural selection, where the strongest and healthiest should survive and reproduce in greater numbers than the weak and sick. This argument—that war results in "negative selection"—had been articulated decades earlier by leading Darwinists, such as Haeckel, as well as leading eugenicists, including Ploetz. For Hitler, then, killing the disabled was one way to overcome the dysgenic effects of modern warfare.[14]

While the war may have increased the urgency for the euthanasia program, Hitler had been persuaded of the benefits of involuntary euthanasia for the disabled for many years. Though he carefully avoided publicly advocating killing the disabled, some of his early rhetoric about the disabled was extremely harsh. In a 1923 interview, while discussing the need to halt the reproduction of the disabled, he quoted a Bible passage that had nothing to do with eugenics to make his harsh views seem religiously justified:

> If thy right eye offend thee, pluck it out, and cast it from thee; for it is profitable for thee that one of thy members should perish, and not that thy whole body should be cast into hell. And if thy right hand offend thee, cut it off and cast it from thee; for it is profitable for thee that one of thy members should perish, and not that thy whole body should be cast into hell.[15]

Thereafter Hitler remarked, "The preservation of a nation is more important than the preservation of its unfortunates." Hitler thus implied that the lives of the disabled do not have to be preserved, if their continued existence conflicts with the interests of the nation. However, at this time he stopped short of overtly promoting killing the disabled.

In 1928 Hitler spelled out the implications of his view of evolutionary struggle for the right to life. He swept away all humanitarian considerations, insisting that death brings progress. He stated,

> Life, however, is struggle. In the struggle for nourishment the one dies, the other lives, and Clausewitz is right to say, "The father of all things is struggle."...Humans have become lord of other living organisms through struggle, for the earth knows no humanitarianism in life today. Does one have a right to something? There are two conceptions of right: your right and the right of nature. The weaker must die, the earth is only for the healthy, and only they have the right to life. In the moment that a *Volk* is defeated, it has received its right, for struggle is the foundation of all dying, but also—all progress.[16]

Here Hitler clearly expressed his belief that the evolutionary struggle should eliminate all humanitarian considerations, including the conception of a natural right to life for all humans, which was a fundamental element of Western human rights philosophy. Further, Hitler's assertion in this speech that only the healthy have the right to life was a disastrous blow to the dignity of the disabled and the sick. In the name of the laws of nature, it overturned centuries of Christian ethical teaching. Hitler thought the rights of the individual were completely subservient to the laws of nature. He also made clear that the death of individuals and even peoples is a necessary part of this struggle. However, because death served a higher purpose—evolutionary progress—Hitler saw the laws of nature, even if they seem harsh and brutal, as ultimately beneficial. This philosophy of biological advance through the death of multitudes took the sting out of death and opened the way for the full-scale slaughter of fellow human beings dismissively defined as "inferior."

By 1928, if not earlier, Hitler seems to have embraced involuntary euthanasia for the disabled. He at least implied this in his *Second Book*. He contrasted the processes of nature, whereby the healthiest survive the struggle for life, with modern society, where people limit births and preserve everyone, "regardless of their true values and inner quality."

He then argued that if one wanted to limit the population, but still preserve its quality, one must keep the birthrate high, but only allow some of the progeny to live. This, Hitler intoned, was the wise racial policy of the Spartans:

> The abandonment of sick, frail, deformed children—in other words, their destruction—demonstrated greater human dignity and was in reality a thousand times more humane than the pathetic insanity of our time, which attempts to preserve the lives of the sickest subjects—at any price—while taking the lives of a hundred thousand healthy children through a decrease in the birth rate or through abortifacient agents, subsequently breeding a race of degenerates burdened with illness.[17]

While Hitler did not expressly state here that he wanted to introduce infanticide as state policy, he did overtly praise the Spartans for killing their weak and sickly infants. He certainly implied that he supported infanticide for the disabled. Hitler's allusion to the Spartans shows the influence—either directly or indirectly—of Haeckel, who had made this argument almost sixty years earlier.

Since Hitler's 1928 manuscript remained unpublished during his lifetime, his contemporaries could not have known about his views expressed therein. However, in a major speech at the Nuremberg Party Congress in August 1929, Hitler also strongly implied that he supported killing the disabled. In that speech he declared,

> If annually Germany would produce a million children and dispose of 700,000 to 800,000 of the weakest, then in the end the result would possibly even be an increase in strength. The dangerous thing is, that we ourselves interrupt the process of natural selection and thereby slowly deprive ourselves of the possibility to increase our population.

Hitler then once again praised Sparta—whom he and his contemporaries understood as practitioners of infanticide—as the "clearest racial state in history." Further, he complained that presently "degenerates are artificially pampered with great effort."[18] Once again, Hitler did not clearly state that he supported infanticide, but what else could it mean to "dispose of" 70 to 80 percent of the children born in Germany. This is an extremely radical proposal that implies mass death. Even though he probably did not intend these numbers to be interpreted as a serious policy proposal, his statement still reflects his general attitude that Germany needs to rid itself of its disabled, weak, and sick members. In both the *Second Book* and in his Nuremberg speech, Hitler's support for infanticide was linked to concern about modern society setting aside the allegedly beneficial effects of natural selection. Killing the disabled would help improve the German people biologically.

One of the most remarkable accounts we have about Hitler's attitudes toward infanticide of the disabled come from the memoirs of a leader in the Nazi SA, Otto Wagener, who had close contact with Hitler before losing favor in mid-1933. Wagener recalled a conversation with Hitler in the summer of 1931, wherein Hitler discussed his enthusiasm for eugenics. According to Wagener, Hitler stated,

> Everywhere in life only a process of selection can prevail. Among the animals, among plants, wherever observations have been made, basically the stronger, the better survives. The simpler life forms have no written constitution. Selection therefore runs a natural course. As Darwin correctly proved: the choice is not made by some agency—nature chooses.[19]

In addition to selection from the outside, however, animal communities also practice a process Hitler called "self-selection." In this process, "Weaklings, runts, sick individuals are cast out of their communities by the healthy ones; some of them are even killed, disposed of. That is the will of nature." Hitler then criticized modern society for not practicing self-selection any more, as the noble Spartans did in ancient times.[20]

In a later conversation that Wagener recalled, Hitler explained that in modern society one should not kill those already living, but should strive to keep the weak and sick from propagating their kind. Whether Hitler really believed this at the time, Wagener naively took him at his word and denied that Hitler could have possibly had anything to do with killing the disabled later on. However, despite Wagener's insistence that Hitler did not intend to kill the disabled, he confessed that Hitler did not regard infants as fully human. Immediately after assuring his entourage that those already living should not be killed, Hitler proclaimed that physicians he had consulted told him "that when a child is born, it is not really fully matured...But if that is so, then the infant does not actually take its place in human society until several months *after* its birth."[21] This view that physicians imparted to Hitler was remarkably similar to Haeckel's view that babies are not yet fully human, which he based on his evolutionary recapitulation theory.[22]

From his own report of Hitler's statements, Wagener should have known that Hitler intended more than sterilization for the disabled.

Indeed a critic of Nazism, the socialist physician Julius Moses, recognized before the Nazis came to power that their program was inherently murderous and that they would enlist physicians as accomplices. In a 1932 article he wrote,

> Everything which until now has been seen as an ethical and moral law, a categorical imperative, for the medical profession, would be thrown overboard by them [National Socialists] like a dirty rag. . . . In the National Socialist "Third Reich" the physician would have the assignment to create a "new and noble human race": only the curable will be healed! The incurably ill are but "useless ballast," "human rubbish," "not worthy to live" and "unproductive." They must be destroyed, utterly destroyed. . . . And it is the physician who must carry out this extermination. In other words, he is to be the executioner.[23]

In 1932 many may have dismissed Moses' analysis as hysterical anti-Nazi propaganda from a Jewish physician, but in retrospect he was simply taking the many inflammatory statements of leading Nazis—including Hitler—seriously. Erich Hilgenfeldt, the leader of the National Socialist Welfare Organization, for example, stated in a 1933 article that "the unfit must be ruthlessly exterminated."[24] By the 1940s it became clear that Moses—who died in a Nazi concentration camp—had accurately depicted the Nazi attitude and policy toward the disabled.

When did Hitler actually decide to kill the disabled? Brandt claimed that before Hitler came to power he already favored a "euthanasia" program. Gerhard Wagner, leader of the Nazi Physicians' League, testified that he approached Hitler at the Nuremberg Party Rally in 1935 about initiating a "euthanasia" program.[25] Though agreeing with Wagner in principle, Hitler urged caution. He thought it would be best to wait until the outbreak of war to initiate it.[26] However, he did not quite wait until World War II began. In early 1939 the parents of a severely disabled baby wrote to Hitler to request permission to have their newborn baby killed. Hitler placed such importance on this request that he sent Brandt to Leipzig with instructions to allow the physicians to kill the child if he concurred with the physicians' diagnosis. He did, and the baby was killed in July 1939. After this initial case, Hitler instructed Brandt and Bouhler to sanction infanticide for other disabled babies.[27]

In the summer of 1939 Hitler also met with Leonardo Conti, who soon became Reich Physician Leader, and other Nazi leaders to initiate an adult "euthanasia" program. At that meeting Hitler said "that he considered it appropriate that life unfit for living of severely insane patients should be ended by intervention that would result in death." Though Conti was willing to take on this assignment, Bouhler convinced Hitler that he and Brandt should organize it instead, since they were already involved with the infanticide program.[28]

Bouhler and Brandt had no difficulty finding physicians willing and even enthusiastic to participate in killing the disabled, for quite a few leading physicians had already jettisoned the idea that the disabled have a right to live. In the 1930s the famous physiologist Emil Abderhalden not only applauded the Nazi's eugenics policies, but he promoted the euthanasia ideology in his medical ethics journal, too. In his analysis of Abderhalden's journal, Andreas Frewer exposes the social Darwinist worldview driving Abderhalden to these positions.[29] Though Abderhalden did not participate directly in the T-4 program, many leading German physicians did. Paul Nitsche had been a respected psychiatrist for decades before the Nazis came to power. He enthusiastically supported euthanasia, and when he became head of the T-4 program in 1941, he recruited like-minded psychiatrists, professors, and directors of asylums.[30] With the outbreak of World War II the program expanded radically, ultimately killing about one out of every four-hundred Germans.

In addition to killing the disabled, Hitler's concern about "negative selection" also drove him to dramatically increase capital punishment during World War II. On several occasions Hitler privately complained that criminals—whom he considered biologically degenerate—were sitting safely in prison, while brave soldiers were giving their lives in battle. In September 1942 he publicly told his fellow Germans that capital punishment would be stepped up during the war, not only to reduce crime, but also to make sure that criminals did not survive the war while German soldiers were giving their lives for the nation.[31] He told his associates the previous month that those with asocial traits should not be tolerated but eliminated, as animals do in their social organizations. Later that month he stated, "If I on the other hand do not ruthlessly eradicate the rabble [of criminals], then one day a crisis will emerge."[32] Hitler was convinced that biologically degenerate criminal elements had fomented or at least participated in the 1918 German revolution, undermining the war effort and bringing on defeat.[33]

Slaughtering Jews

As we have already seen, Darwin, Haeckel, and their followers believed that human races were locked in a struggle for existence that would decide which races would survive and which would perish. In *Descent of Man* Darwin predicted that the advanced races (that is, Europeans) would ultimately exterminate the primitive races, such as the black Africans and American Indians. Haeckel and many German Darwinists argued likewise in many of their writings. Darwin's view of racial extinction was shaped in part by the recognition that Europeans were taking over vast stretches of territory in the Americas and Australia at the expense of indigenous peoples there.[34] Many scholars have noted the way that social Darwinism both adopted colonialist racism and then promoted it as scientific.[35] Though Darwin at times expressed sympathy for those of other races, he also exulted in the European triumph. Late in life he wrote to a colleague that the "more civilised so-called Caucasian races have beaten the Turkish hollow in the struggle for existence. Looking to the world at no very distant date, what an endless number of the lower races will have been eliminated by the higher civilised races throughout the world."[36]

Darwin never advocated killing people based on their racial character, nor did he look upon the Jews as an inferior race needing to be eliminated, as Hitler did. However, his views on racial extermination did help shape social Darwinist discourse in Germany, which was often more callous and more belligerent toward other races than Darwin's own views.[37] Hitler would adopt these social Darwinist ideas about racial extermination and blend in his hatred of Jews. Stig Förster and Myriam Gesslerr have recently argued that Nazi genocide would not have emerged without social Darwinism, racism, and "a perverse analysis of World War I," all three of which "lay at the core of Nazi ideology."[38] Many other historians have noticed the importance of social Darwinism in driving Nazi genocide.

Also, as the historians Henry Friedlander and Robert Proctor have recognized, the Nazi campaign to annihilate the Jews was in many respects a continuation of the T-4 program.[39] Ideologically the two policies were linked, since both aimed at exterminating those deemed biologically inferior. In practice the two were connected, since when the T-4 killing centers closed in August 1941 many of the participating personnel were shipped east to the new death camps being created in occupied Polish territory. The T-4 program had served as a testing ground for killing methods, and each "euthanasia" facility constructed gas chambers as the most efficient method for mass killing.

As we have already seen (chapters 3–4), Hitler often portrayed his campaign against the Jews as part of the struggle for existence between the allegedly superior Aryan race and the supposedly inferior Jews. In a dinner monologue in April 1942, while Jews were being slaughtered in death camps, he ominously stated that "one may not have any pity for people who have been fixed by destiny to perish." Hitler's audience probably did not understand the implications of this statement, since Hitler was at the time discussing the fall of ruling classes in the past. However, he continued by stating that "pity must of course be limited and must restrict itself to the members of one's own nation." He then explained how this view fit into his own evolutionary framework: "As in all domains, so also in the area of selection, nature is the best teacher. One could not think up a better design for nature than the advance of organisms through it [selection]: only in hard struggle." He immediately connected these evolutionary ideas to Jews, complaining about those who showed pity for them, because pity should only be reserved for fellow Germans.[40] Pitilessly murdering the Jews, then, was part of a program intended to bring about "the advance of organisms" through struggle and selection.

This does not mean that Hitler had coherent plans for killing all European Jews long before 1941. I am persuaded by the historical evidence that Christopher Browning, Saul Friedländer, and many other historians are right to argue that Hitler did not have fixed plans to kill all the European Jews until sometime in the second half of 1941 (I do not take a position in the debate over exactly when in 1941 he decided on this). Hitler's social Darwinist vision of a racial struggle-to-the-death did not necessarily mean that Germans had to shoot or gas Jews. Deportation of Jews from German territories would achieve his goals of racial purity and expanded living space perfectly well.

However, even though Hitler did not necessarily have to commit genocide to accomplish the goals implicit in his evolutionary ethic, his long-term vision was in many respects genocidal. Since he thought the Aryan race was biologically superior to all other races, he ultimately expected them to thrive, reproduce abundantly, and ultimately outstrip all other races. In Hitler's view of long-term historical development, all other races were slated for extinction as the Aryans gradually gained more and more living space at their expense. This could not be accomplished in one or two generations, but Hitler was determined to make as much "progress" as he could.

In Hitler's earliest writing on anti-Semitism in 1919, he advocated removing Jews from Germany. He often used the words "eliminate," "remove," and "get rid of" when referring to the Jews, which most people would have construed as forced emigration, especially since that was the message conveyed by the official Nazi Party Program. However, in his earliest political speeches Hitler did at times use more inflammatory and even murderous language in his harangues against the Jews. In notes for a speech in 1921 he wrote, "Hang the Jews, hang the profiteers, racial combat."[41] The following year he noted that Germans were involved in a racial struggle, not a class struggle. Then he ominously stated, "Struggle means to destroy."[42] Since these are merely notes for speeches, it is not clear exactly how he articulated these points in actually delivering the speeches. His intended rhetoric seems murderous nonetheless.

However, in a 1920 speech Hitler showed his ability to use ambiguity to imply murderous intentions without actually overtly advocating murder. He stated,

> Do not think that you can fight against an illness without killing the germ, without destroying the bacillus; and do not think that you can fight against the racial tuberculosis without making sure that the *Volk* gets free from the germ of the racial tuberculosis. The effects of Jewry will never perish, and the poisoning of the *Volk* will not end, as long as the germ, the Jew, is not removed from our midst.[43]

By using the trope of the Jew as a harmful microorganism and noting that microorganisms must be killed or destroyed, he implied that Jews should be killed, too. However, overtly he only called for removing Jews, implying that deportation would suffice.

One accusation Hitler regularly hurled at the Jews was conspiring to destroy the racial health and vitality of the German people. In his famous 1920 speech, "Why Are We Anti-Semites?" he claimed that the Jews were promoting racial degeneration among the German people, because they feared a robust and healthy German people.[44] In a 1924 article, Hitler wrote that the Jews were using Marxism to destroy the racial foundations of Germany.[45] Hitler often blamed Marxism—which he saw as one prong of the Jewish conspiracy for world dominion—for weakening Germany's racial vitality by teaching human equality, rather than recognizing the biological doctrine of the inequality of individuals and races. In a 1936 speech he censured communism for its alleged aim: "to exterminate what is healthy, what is healthiest of all, in fact, and to place in its stead the most degenerated of all."[46] Hitler's opposition to the Jews, then, was not just because they were allegedly immoral or because they were taking advantage of Germans economically—which he thought they were—but also because he thought they were consciously sponsoring biological decline among Germany's population.

In the six years of Nazi rule before World War II, policy toward Jews intensified from discrimination to exclusion to deportation. In November 1938, when the Nazi regime orchestrated the Crystal Night pogrom, the first nationwide violence against the Jews, the regime specifically ordered that Jews not be killed (though some were murdered, anyway). Thousands of male Jews were rounded up and sent to concentration camps, but most were released within a few weeks and warned to get out of Germany. In the ensuing months Jewish businesses and property were confiscated, forcing many Jews to emigrate. From 1933 to the outbreak of World War II about 300,000 of Germany's Jewish population had fled. Only 200,000 remained. Furthermore, when Germany annexed Austria in early 1938, the SS sent Adolf Eichmann to Vienna to organize the deportation of Jews from Austria. Deportation was official Nazi policy in 1938–1939.

In January 1939, while preparing for military action against Czechoslovakia and Poland, Hitler uttered his famous "prophetic" warning against the Jews. At the sixth anniversary of taking power in Germany, Hitler proclaimed, "Should the international Jewry of finance (*Finanzjudentum*) succeed, both within and beyond Europe, in plunging mankind into yet another world war, then the result will not be a Bolshevization of the earth and the victory of Jewry, but the annihilation (*Vernichtung*) of the Jewish race in Europe."[47] It seems clear that the aim of this "prophecy" was to warn Britain, France, the United States, and the Soviet Union from interfering with his expansionism. It was not an indication that Hitler had already decided to annihilate Jews. Hitler mentioned this prophecy often thereafter, even in private conversations with Nazi leaders, so he seemed to take it seriously. Some scholars take his statement so literally that they think Hitler waited until the United States was in the war—making it a real world war—before deciding to systematically exterminate all the Jews in German control.[48] However, even if Hitler's decision came earlier, his prophecy was contingent on a world war.

When Germany invaded Poland in September 1939, they gained control over an additional three million Polish Jews. They also provoked

war with Britain and France. However, even thereafter the SS continued planning the deportation of Jews. Though some Jews had been
killed during the invasion of Poland, the vast majority had been isolated in ghettoes. In May 1940 the SS concocted a plan to ship these
Jews to Madagascar, a French colony, once France was defeated. Hitler
approved of the Madagascar Plan, for during a private conversation in
August 1940 he stated, "Later we want to ship the Jews to Madagascar.
There they can even build up their own state."[49] Many scholars believe
that if the Nazis had succeeded in resettling the Jews to Madagascar,
it would have resembled a giant concentration camp. However, even
if Hitler was sincere about allowing the Jews to build their own state
on Madagascar, his intent was still destructive. Hitler had continually
argued that Jews were parasitic on other states and that they did not
have the requisite moral instincts to found their own state. In *Mein
Kampf* he wrote that if Jews were forced to live among themselves,
they would "turn into a horde of rats, fighting bloodily among themselves." They would engage "in hate-filled struggle and exterminate
one another."[50]

Any plan for deporting Jews was, in Hitler's mind, simply a way to
destroy the Jews, sooner or later. If he deported them to a Jewish colony, he thought they would destroy themselves. If he deported them to
other countries, he was confident that anti-Semitism would increase in
those countries, ultimately undermining the Jews' position there. Or, if
the Jews somehow gained power in other countries, as he thought they
had in Bolshevik Russia, they would only help Germany by weakening those nations. He told Goebbels once that he was determined not
to send the Jews to Siberia, because he feared the harsh climate would
help the Jews develop into a stronger race (just as he thought the harsh
northern climate had brought biological improvement to the Nordic
race in primeval times).[51] In Hitler's thinking, then, deportation was
ultimately a method to destroy the Jews.[52]

When Germany invaded the Soviet Union in June 1941, Hitler's
campaign against the Jews stepped up in intensity. Special SS and police
units were sent into the conquered territories of the Soviet Union to
kill Jews and communist functionaries. There is still no convincing
evidence that Hitler intended to kill all the Jews of Europe in June
1941, especially since at first the SS units did not even kill all the Soviet
Jews they accosted. Sometime in the last half of 1941, however, Hitler
decided to make good on his prophecy and annihilate all the Jews
within his grasp. In August 1941 he told Goebbels that his prophecy
was being confirmed, since "In the East the Jews are having to pay

the bill."[53] This was still not a clear statement that all European Jews were targeted for destruction, though it might have implied it. Only in October 1941 did the Nazi regime forbid Jews from emigrating from German-occupied territory, which might mean that the decision for their annihilation had already been made.[54] On December 13, 1941, Hitler made an unequivocal statement to Nazi leaders about the annihilation of the Jews. Goebbels recorded Hitler's comments in his diary:

> Concerning the Jewish question the Führer is determined to make a clean sweep (*reinen Tisch zu machen*). He prophesied to the Jews, that, if they would cause a world war again, they would then experience their destruction. That was no hollow phrase. The world war is here, so the destruction of Jewry must be the necessary consequence. This question is to be viewed without any sentimentality. We are not here to have pity on the Jews, but only pity on our own German *Volk*.[55]

Soon thereafter, Jews from all over Europe were being sent to the death camps in former Polish territories.

The war, then, radicalized Hitler's policies, though his primary goals never altered. He was still determined to gain living space and to win the racial struggle for existence. In a private speech to generals and officers in June 1944, Hitler spoke extensively about the German need for more living space. He presented the war as a part of the natural process of "natural selection of the stronger and simultaneously the process of getting rid of the weaker." He admitted that the process seems cruel, but he assured these officers that it was really wise and, in any case, inescapable. At the close of this speech he informed them that he had "eliminated" the Jews as a "foreign racial element" to make space for hundreds of thousands of fit Germans.[56]

In a private speech a month earlier Hitler had made even clearer the evolutionary reasoning that motivated his extermination of the Jews. He justified his intolerance toward others by appealing to evolution. Nature is the most intolerant thing around, because it gets rid of every weak being. "It destroys everything that is not entirely capable of life, that will not or cannot defend itself." It may seem brutal to us that the female dog pushes aside the runt, but this is really a wise move, Hitler stated, presumably because it promotes biological vitality. After laying this groundwork, Hitler mentioned that some people wonder why it is necessary to be so harsh toward the Jews. He answered: "I have pushed the Jews out of their positions, indeed ruthlessly pushed them out. Here

I have acted, just as nature does, not brutally, but rather according to reason, in order to preserve the better ones [that is, Aryans], and I have thereby freed up hundreds of thousands of positions." These positions were now open for good German children, he promised. Getting rid of the "inferior" Jews to make room for the "superior" Germans was—in Hitler's view—part of the natural evolutionary process. Hitler then made clear that this principle defined his ethic, stating, "For here also we recognize only one principle, namely the preservation of our race, preservation of our species. Everything that serves this principle is right. Everything that is detrimental to it is false."[57] In this speech late in the war Hitler justified killing the Jews by appealing to his evolutionary ethic.

Leading German scientists and physicians supported and assisted Hitler in his drive to eliminate the Jews. The world famous geneticist Lenz wrote in the 1936 edition of the major text he coauthored that Jews were a harmful "race of parasites," and organisms "thrive better without parasites." Lenz was a convinced anti-Semite long before the Nazis came to power, and he had integrated anti-Semitic elements into his earlier writings. In 1943, while the Jews were being massacred, Rüdin praised Nazi policies, including "the fight against parasitic foreign-blooded races, such as the Jews and Gypsies."[58] Many German anthropologists cooperated with the Nazi regime, which they believed was implementing their racial agenda.[59] Some physicians even participated directly. At Auschwitz a respected colleague and student of the leading Nazi eugenicist Otmar von Verschuer, Joseph Mengele, who held two doctorates (physical anthropology and medicine), determined which Jews would live and which would die. He even sent tissue samples from corpses in Auschwitz to Verschuer.[60]

In sum, Hitler's evolutionary ethic did not require killing. He could have merely sterilized the disabled and deported the Jews. This would have accomplished his goals of expanding the German population, strengthening the Aryan race by eliminating "inferior" individuals and races, and expanding German living space. However, even though killing may not have been required by Hitler's evolutionary ethic, Darwinism contributed nonetheless to the death of the disabled and Jews. Christopher Hutton is right when he asserts in the closing sentences of his book on Nazi racism:

All the key elements of this [Nazi] world-view had been constructed and repeatedly reaffirmed by linguists, racial anthropologists, evolutionary scientists and geneticists. Ludwig Plate [a Darwinian biologist at the University of Jena] observed that "progress in evolution goes forward over millions of dead bodies" (Plate 1932:vii). For Nazism, survival in evolution required the genocide of the Jews.[61]

As Darwinists consistently taught, the struggle for existence necessarily resulted in mass death for the "unfit," which caused evolutionary progress. Hitler—along with some other Darwinists—believed that the right to life only belonged to the "fit," which they interpreted as the healthy and strong. Those who were "unfit"—whether disabled individuals or those deemed racially inferior—were slated for destruction anyway. By killing the "inferior" he thought he was merely restoring the balance of nature and fostering evolutionary progress.

Conclusion

I do not know how many times I have heard colleagues and friends ask if the title of my book is an oxymoron. How could a mass murderer like Hitler have had any ethic? Yet, surprisingly, the fanaticism that motivated him to pursue mass killing—and many other policies—stemmed at least in part from his sincerely held (but pernicious) conviction that killing people he deemed inferior would serve a higher moral purpose: advancing the human species in the evolutionary process. This kind of evolutionary ethics was central to Hitler's ideology and practice, because ultimately Hitler measured every policy by its effect on biological improvement. Various Nazi leaders, such as Rudolf Hess and Hans Schemm, agreed with the geneticist Fritz Lenz that Nazism was "applied biology."[1]

We have already examined a multitude of statements by Hitler in *Mein Kampf*, his *Second Book*, his speeches, and in private conversations, where he expressed his views on evolutionary ethics. However, Hitler's thought was also reflected in the Nazi booklet, *Why Are We Fighting?*, which explained that the Nazi war effort was part of an ideological struggle. The preface of this work reproduced Hitler's January 8, 1944, decree directed to all military officers, in which he wrote, "I command therefore, that the worldview contained in this book be instilled convincingly and emphatically in the soldiers in regular instruction."[2] Hitler made clear in this decree that he personally endorsed the ideology expressed in this booklet.

Evolutionary ethics permeates *Why Are We Fighting?*; it also stresses the centrality of biological racism and eugenics for the Nazi worldview. The anonymous author(s) approvingly quoted a German eugenicist that "the natural laws, according to which the cosmos of dying and becoming transforms itself and evolves, are divine laws." This point is reiterated again later in the booklet: "Our racial idea is only the 'expression of a worldview,' which recognizes the higher evolution of humans as a divine law." This notion that the chief article of the Nazi worldview is promoting the higher evolution of humanity is expressed repeatedly throughout the booklet in many different ways. True socialism, it stated, "means preservation and further evolution of the *Volk* on the basis of its characteristic laws of evolution." This booklet also stressed the importance of race in the evolutionary process:

> Thus we believe *in the task of the elevation of humans*. Ultimately our struggle serves this purpose, and our struggle must be inexorable against everything that opposes this task, for the appropriate fulfillment of this task is dependent on the highest-evolved, most creative, and most capable race maintaining its decisive influence on the living arrangements of the peoples of the earth.[3]

Racism, as central as it was in the Nazi worldview, was important to Hitler and other Nazis because it contributed to the improvement of humanity in the evolutionary process.

Hitler considered the Aryan or Nordic race the highest form of humanity, and he thought the German people were predominantly Aryans. Thus, anything that would help them outreproduce all the other "inferior" races would ultimately lead to biological progress for the human species, he reasoned. Not only did he consider the Aryans superior mentally, but he also thought they were superior morally. As a biological determinist, he—like many scientists of his day—believed that morality was biologically innate and hereditary. Thus, the triumph of the Germans in the racial struggle for existence would not only lead to an advance in culture, but it would also lead to moral improvement.

Hitler hoped to drive evolution forward by clearing away inferior races, making room for his beloved fellow-Germans. Many Nazi policies were aimed at hindering the reproduction of races deemed inferior. Jews were a special target of Nazi discrimination and persecution, because Hitler supposed that Jews were biologically immoral. Eliminating Jews—one way or another—would rid the world of immorality.

Racist evolutionary ethics underpinned, influenced, or justified many important features of Nazi ideology and practice. Even ideological currents that predated Darwinism by centuries, such as anti-Semitism, were recast and reshaped by evolutionary thinkers who influenced Hitler's thought. Evolutionary ethics shaped Hitler's view of history as a racial struggle for existence, as well as his insistence on competition among Germans for political and economic positions. Nazi pro-natalist and eugenics policies aimed at improving the human species by increasing the reproduction of the "superior" racial elements, while limiting procreation of the "inferior" ones. The T-4 "euthanasia" program and the attempt to annihilate the Jews were radical measures to get rid of those deemed "inferior." The drive for living space was also built on social Darwinist principles, since the goal was to provide land and resources for more German settlers at the expense of racial inferiors.

Of course, evolutionary ethics does not explain everything about Hitler's ideology. Hitler was syncretistic in building his ideology, drawing on many different currents of thought. Not only was he influenced by many famous Germans, such as Wagner, Schopenhauer, and Frederick the Great, but he imbibed many ideas from the press, periodicals, and from his own experiences, especially during and after World War I. He integrated elements from disparate sources, such as Pan-German nationalism, Christian anti-Semitism, Prussian militarism, the Nietzschean will to power, and many others in constructing his worldview. He, like all men, was also motivated by many noncognitive factors, such as fear, pride, and covetousness.

Social Darwinists, both before and during the Nazi period, also integrated many preexisting ideas into their ideology. Racism obviously preexisted Darwinism, so it was not derived from evolutionary ethics. However, Darwin and other Darwinists—especially Haeckel, Woltmann, Lenz, and Fischer—integrated racism into evolutionary theory. They explained that the "inferior" races had not evolved as far from their simian ancestors as the more highly evolved Europeans.

So, while evolutionary ethics does not even come close to explaining the origin of all Nazi ideology, it was nonetheless a central element influencing many facets of Nazi ideology, especially pro-natalism, eugenics, offensive warfare to gain living space, killing the disabled, and racial extermination. Even when Nazi policies were inconsistent and contradictory, often the primary goal was immutable. Hitler was willing to compromise on details, but only if he thought it would advance the highest purpose of life. Thus, by identifying the core goal of Hitler's ideology—evolutionary progress—we can nuance the claim that Nazi policies were essentially pragmatic and opportunistic. Sometimes apparent opportunism flowed from the Nazis' own confusion about how best to attain evolutionary progress. I also do not deny that in some cases Hitler followed an opportunistic course, especially in relation to the timing of decisions. Sometimes he put off policy decisions he knew were presently unattainable, biding his time until the moment was auspicious for implementing his aims. Thus, even while pursuing pragmatic goals he never lost sight of his ultimate goal of the biological improvement of humanity, which he believed would also bring cultural progress in its train.

One key example of the tension between Nazi ideology and pragmatism was the treatment of conquered Slavs. Though much recent scholarship on Nazi foreign policy and warfare emphasizes the ideological underpinnings of this project, some scholars claim that the details of Nazi policy in Slavic lands were ad hoc, haphazard, and even contradictory.[4] Some of these inconsistencies were the result of problems within Nazi ideology. For instance, Nazis could never find a way to scientifically determine who belonged to the Aryan or Nordic race. It was not inconsistent for Nazis to seek Aryan racial comrades among Czechs or Poles, especially those in areas bordering with Germany. However, how to sift through the Slavic populations presented an insurmountable problem, one that the Nazis muddled through.

Another alleged inconsistency in Nazi policy toward Slavs was the shift from deporting Poles to exploiting them as slave labor.[5] However, in this case, it is not clear that this was really an inconsistency. Several months before the Polish campaign, Hitler secretly told his military leaders that, since the non-German populations in occupied territories would not be called up for military service, they "are available for labor service."[6] Deportation and forced labor were both options available to the Nazi regime to gain their far-reaching goals. Ultimately, deportations would be necessary for Germans to settle their coveted living space, but since there simply were not enough Germans and fellow Aryans available to populate all the conquered territories, enslaving the Slavs was the intermediate step. Uwe Mai is correct when he asserts, "The Darwinian orientation of the [Nazi] conception of settlement [of occupied territory] proceeded from the view that the superior German settler would displace the racially inferior people in a process guided by natural laws. Their own task was to accelerate and consolidate this process."[7] The Nazis did not always have coherent plans about how to accomplish this resettlement, but whatever policies they pursued, they always kept their primary ideological goal of racial improvement in mind.

While it is not at all clear how successful the Nazis were at imbuing all their fellow-Germans with their racist version of evolutionary ethics, it is obvious they tried. Many avenues of Nazi propaganda spread this new Nazi gospel. Hitler preached it in his speeches and in *Mein*

Kampf. It was a central feature of his chapter on "Race and Nation," which was the only chapter from *Mein Kampf* mass-produced as a pamphlet during the Third Reich. Gilmer Blackburn, who has carefully examined Nazi school textbooks for their views about race and history, argues that "the bedrock of the Nazi conception of life was certainly the Darwinian struggle for existence." Blackburn further stated, "Hitler's view of nature subordinated the harmony and the nobility of the Hegelian view of the world to the hideous and cruel characteristics of the Darwinian view."[8] In 1933 the Nazi education minister decreed that biological thought would be the foundation for instruction in all subjects, especially German language, history, and geology.[9] The biology curriculum was heavily imbued with Darwinian evolution and racism, but other subjects included heavy doses of social Darwinism, too.[10]

Many other works sponsored by the Nazi regime promoted evolutionary ethics and social Darwinism to the masses. Pamphlets reflecting evolutionary ethics, such as *Why Are We Fighting?* and *Racial Policy* were circulated en masse and used in educational efforts in the military and in schools.[11] Many periodicals were either founded or co-opted by the Nazis to disseminate their evolutionary ethic. Emil Abderhalden's journal *Ethics* reached an audience of scientists and especially physicians with social Darwinist and eugenics views paralleling Hitler's own ethical views.[12] Other scientific journals devoted to eugenics, racism, and biology purveyed Nazi ideals to the educated elites. The glossy magazine, *Neues Volk*, published by Walter Gross's Racial Policy Office, reached a more popular audience with articles about eugenics, racism, population expansion, and other ideas emanating from Hitler's racist evolutionary ethics. The Nazi regime also produced feature films and documentaries promoting social Darwinism, eugenics, and euthanasia to reach both popular and scholarly audiences.[13] On the basis of his examination of Nazi films used in medical education, Ulf Schmidt concludes that "Nazism reveals a fundamental break with Judaeo-Christian ethics, an attack on a traditional belief system based on altruism and compassion." Rather, he explains, these films were permeated with the social Darwinist struggle for existence.[14]

But are these ideas about social Darwinism, evolutionary ethics, eugenics, and scientific racism that I have discussed in this book really scientific? Are they not just pseudoscientific justifications for following prejudices and irrational ideas, as many scholars have insisted?[15] In the sense that many of Hitler's ideas, including his vaunted "scientific racism," were empirically false, of course his views were pseudoscientific.

Many of these ideas were shaped by prejudices and irrational beliefs rather than empirical data, so in this sense they were certainly pseudoscientific. Most scientists, especially after the Nazi period, repudiated scientific racism and eugenics, exposing many of their foundational ideas as false and misleading. No reputable scientist today would endorse Hitler's views as scientific.

However, recently many historians of science have expressed discomfort about assigning the appellation pseudoscience to Nazi racial thought, because calling it pseudoscientific is anachronistic. Benno Müller-Hill, for instance, while decisively rejecting Nazi eugenics and racism, nonetheless forcefully denies that Nazi racism and eugenics should be labeled pseudoscience. Müller-Hill, along with many other historians of science, defines science as what most scientists accept as valid at any given time, even if later those ideas are shown to be mistaken. Using this definition, many elements of Nazi racism were not pseudoscientific, because many—perhaps most—biologists, anthropologists, and medical professors accepted scientific racism as valid science.[16]

For my purposes, the important point is not whether we label Hitler's evolutionary ethic scientific or pseudoscientific. (Today the battle over the status of some forms of evolutionary ethics—though not the racist form that Hitler embraced—is still raging; many sociobiologists and evolutionary psychologists assert it is scientific, while others consider it—or at least some aspects of it—"junk science.") The important point to recognize is that evolutionary ethics, scientific racism, eugenics, and many related ideas were considered mainstream scientific ideas before and during the Nazi regime, even though they were contested. Many leading scientists and physicians—and not just in Germany—believed that morality is the result of evolution, that behavior was determined by one's hereditary characteristics, and that some races had higher ethical tendencies than others. Most leading anthropologists in Germany by the early twentieth century believed that Europeans were mentally and morally superior to "inferior" races, such as black Africans, American Indians, or Australian aborigines. Some included Jews in the category of hereditarily inferior races. Eugenics was accepted by large segments of the medical community, too, spanning the political spectrum from left to right.

The acceptance by scientists and physicians of many elements making up Hitler's worldview helps explain the ready acquiescence and even eager participation of many highly educated Germans in Nazi atrocities. Like Hitler, many of them believed that humans should seize control of the evolutionary process to foster biological improvement. Some would have been (and were) horrified by the radical measures introduced by Hitler and his regime to accomplish these goals, such as killing the disabled, annihilating "inferior" races, and launching offensive wars. However, incredibly, quite a few did not balk when asked to assist the Nazi regime carry out its worst atrocities. Some scientists and physicians not only cooperated with the Nazi compulsory sterilization program for the disabled and half-blacks, but they also supported and participated in killing the disabled, Jews, and Gypsies. Scholars, especially anthropologists and physicians, rallied to the Nazi regime's call for experts to racially categorize people, so they could determine their fitness for surviving, marrying, and/or reproducing. Life and death was in the hands of these scientists and physicians, who sometimes even went beyond the regime's directives in committing atrocities. Most of them apparently thought—just as Hitler did—that their actions were morally justified, since they were contributing to the progress of the human species.[17]

The purpose of my analysis of Hitler's ethic is by no means to exonerate him for his crimes against humanity by explaining that he really had "good intentions." On the contrary, the point is that evil can be—and often is—perpetrated under the guise of doing good. Hitler—and other fanatical utopians—erred morally by believing that his vision of a better future for humanity should be imposed regardless of the present human cost. He dispensed with any fixed morality that interfered with his "higher" goals. For him the ends justified the means. Though their ideologies were in some respects polar opposites, Hitler and Lenin both considered it morally justified to ride roughshod over any people who interfered with their vision of historical development. Racial enemies under Hitler and class enemies under Lenin or Stalin were persecuted in the quest for a higher humanity that would produce a higher culture.[18] While trying to create a better world of higher humans building an advanced culture with greater morality, Hitler plunged into the abyss instead.

NOTES

Introduction

1. Ron Rosenbaum, *Explaining Hitler* (New York: Random House, 1998).
2. Robert S. Wistrich, *Hitler and the Holocaust* (New York: Modern Library, 2003), 7, 240.
3. One major exception is Claudia Koonz, *The Nazi Conscience* (Cambridge, MA: Harvard University Press, 2003).
4. My views on this have been influenced and stimulated by Michael Burleigh and Wolfgang Wippermann, *The Racial State: Germany, 1933–1945* (Cambridge: Cambridge University Press, 1991).
5. A few scholars have suggested that Hitler and the Nazis embraced a consistent ethic: Koonz, *Nazi Conscience*; Peter Haas, *Morality after Auschwitz: The Radical Challenge of the Nazi Ethic* (Philadelphia: Fortress Press, 1988). Neither identify this ethic as a form of evolutionary ethics. In an essay, however, Peter Haas does astutely identify the Nazi ethic with social Darwinism: "Science and the Determination of the Good," in *Ethics after the Holocaust: Perspectives, Critiques, and Responses*, ed. John Roth (St. Paul: Paragon House, 1999), 49–89; however, Haas does not analyze Hitler's own views in any depth.
6. Richard J. Evans, *Third Reich in Power* (New York: Penguin, 2006), 259; and Hans-Walter Schmuhl, *Rassenhygiene, Nationalsozialismus, Euthanasie. Von der Verhütung zur Vernichtung 'lebensunwerten Lebens' 1890–1945* (Göttingen: Vandenhoek und Ruprecht, 1987), 151, both mention the ethical thrust of social Darwinism. Joachim Fest, *Hitler*, trans. Richard and Clara Winston (New York: Helen and Kurf Wolff, 1974), 205–210, 37, 53–56, 201, 608, claims that Hitler based his ethics on nature and struggle. Many scholars have noted the importance of social Darwinism in Hitler's world view: Ian Kershaw, *Hitler*, 2 vols. (New York: Norton, 1998–2000), 2:xli; see also 1: 290, 2: 19, 208, 405, 780; Richard J. Evans, *The Coming of the Third Reich* (New York: Penguin, 2004), 34–35, and *Third Reich in Power*, 4, 708; Eberhard Jäckel, *Hitler's World View: A Blueprint for Power* (Cambridge, MA: Harvard University Press, 1981), ch. 5; Mike Hawkins, *Social Darwinism in European and American Thought, 1860–1945: Nature as Model and Nature as Threat* (Cambridge: Cambridge University Press, 1997), 277–78; Rainer Zitelmann, *Hitler: Selbstverständnis eines Revolutionärs* (Hamburg: Berg, 1987), 15, 466; Karl Dietrich Bracher, *Die Deutsche Diktatur. Entstehung, Struktur, Folgen des Nationalsozialismus*, 7th ed. (Cologne: Kiepenheuer and Witsch, 1993) 13–15; Gerhard Weinberg, *The Foreign Policy of Hitler's Germany*, vol. 1: *Diplomatic Revolution in Europe, 1933–36* (Chicago: University of Chicago Press, 1970), 1–6; Wolfgang Wippermann, *Der consequente Wahn. Ideologie und Politik Adolf Hitlers* (Gütersloh: Bertelsmann, 1989), 179; Robert Gellately and Nathan Stolzfus, "Social Outsiders and the Construction of the Community of the People," in *Social Outsiders in Nazi Germany*, ed. Robert Gellately and Nathan Stolzfus

(Princeton, NJ: Princeton University Press, 2001), 4; Neil Gregor, *How to Read Hitler* (New York: Norton, 2005), 40; Alan Bullock, *Hitler and Stalin: Parallel Lives* (New York: Alfred Knopf, 1992), 23, 142; Stig Förster and Myriam Gessler, "The Ultimate Horror: Reflections on Total War and Genocide," in *A World at Total War: Global Conflict and the Politics of Destruction, 1937–1945*, ed. Roger Chickering, Stig Förster, and Bernd Greiner (Cambridge: Cambridge University Press, 2005), 67; Hans Staudinger, *The Inner Nazi: A Critical Analysis of Mein Kampf* (Baton Rouge: Louisiana State University Press, 1981), 78–79; Werner Maser, *Adolf Hitler: Legende, Mythos, Wirklichkeit* (Munich: Bechtle, 1971), 168, 236, 255–56, 283–84; Brigitte Hamann, *Hitler's Vienna: A Dictator's Apprenticeship*, trans. Thomas Thornton (New York: Oxford University Press, 1999), 102, 202, 202–203; Jost Hermand, *Old Dreams of a New Reich: Volkish Utopias and National Socialism*, trans. Paul Levesque (Bloomington: Indiana University Press, 1992), 63; Gilmer Blackburn, *Education in the Third Reich: Race and History in Nazi Textbooks* (Albany: State University of New York Press, 1985), 21–22; Edward Westermann, *Hitler's Police Battalions: Enforcing Racial War in the East* (Lawrence: University Press of Kansas, 2005), 58; see also Hans-Günter Zmarzlik, "Der Sozialdarwinismus in Deutschland als geschichtliches Problem," *Vierteljahrshefte für Zeitgeschichte* 11 (1963): 246–273. John Lukacs, *The Hitler of History* (New York: Vintage, 1997), 120–127, is one of only a few scholars to claim that social Darwinism was not very important in Hitler's ideology.

7. My dissertation is published as *Socialist Darwinism: Evolution in German Socialist Thought from Marx to Bernstein* (San Francisco: International Scholars Publications, 1999).

8. Richard Weikart, *From Darwin to Hitler: Evolutionary Ethics, Eugenics, and Racism in Germany* (New York: Palgrave Macmillan, 2004). Though I have constructed this present book so it can stand alone, it is a sequel to *From Darwin to Hitler*, so anyone wanting detailed information about forms of evolutionary ethics in Germany before Hitler should consult my earlier work.

9. Adolf Hitler, *Mein Kampf*, trans. Ralph Manheim (Boston: Houghton Mifflin, 1943), 287–289.

10. Adolf Hitler, "Hitler vor Bauarbeitern in Berchtesgaden über nationalsozialistische Wirtschaftspolitik am 20. Mai 1937," in *"Es spricht der Führer": 7 exemplarische Hitler-Reden*, ed. Hildegard von Kotze and Helmut Krausnick (Gütersloh: Sigbert Mohn Verlag, 1966), 220–221.

11. "Hitler vor Offizieren und Offiziersanwärtern am 15. Februar 1942," in ibid., 306–7.

12. Peter Walkenhorst, *Nation—Volk—Rasse: Radikaler Nationalismus im Deutschen Kaiserreich 1890–1914* (Göttingen: Vandenhoeck and Ruprecht, 2007), shows how "radical nationalists" fused nationalism and racism under the influence of social Darwinism.

13. This is already apparent in Hitler's early writings and speeches; see Hitler to Adolf Gemlich, September 16, 1919; and report of Hitler, "Der Arbeiter im Deutschland der Zukunft," (November 19, 1920), in *Hitler: Sämtliche Aufzeichnungen, 1905–1924*, ed. Eberhard Jäckel (Stuttgart: Deutsche Verlags-Anstalt, 1980), 88–89, 262.

14. The term Aryan is a misnomer, and I use it reluctantly, but since Hitler used it so often, I feel compelled to use it. I considered using scare-quotes around it, but finally decided against it, since this would make the text more cumbersome.

15. Hitler, *Mein Kampf*, 402; emphasis in original. I have substituted the word "evolution" for "development" in this quotation as the proper translation for "*Entwicklung*," and I do so everywhere that the context demands it. "Development" is a proper translation for "*Entwicklung*" in some contexts, but "*Entwicklung*" is also the standard German term for biological evolution. Though Manheim inexplicably never translates Entwicklung as evolution, even in contexts clearly discussing biological transmutation, many translators of Hitler's other writings and speeches, including his *Second Book*, frequently translate "*Entwicklung*" as evolution.

16. "Hitler vor Leitern der Rüstungsindustrie auf dem Obersalzberg Anfang Juli 1944," in *"Es spricht der Führer,"* 336.

17. Charles Darwin, *The Descent of Man*, 2 vols. in 1 (London, 1871; reprint. Princeton: Princeton University Press, 1981), 1: 98.

18. George L. Mosse, *Nazi Culture: Intellectual, Cultural, and Social Life in the Third Reich* (New York: Grosset and Dunlap, 1966), xxvi; George L. Mosse, *Nationalism and Sexuality: Respectability and Abnormal Sexuality in Modern Europe* (New York: Howard Fertig, 1985), ch. 8.

19. Koonz, *Nazi Conscience*, 1–2, 254–255; see also Jonathan Glover, *Humanity: A Moral History of the Twentieth Century* (New Haven: Yale Nota Bene, 2001), 317, 355.

20. Jäckel, *Hitler's Weltanschauung*. Some scholars emphasizing the importance of ideology are Gregor, *How to Read Hitler*; Michael Burleigh, *The Third Reich: A New History* (New York: Hill and Wang, 2000); Evans, *Coming of the Third Reich*; Evans, *Third Reich in Power*; Peter Fritzsche, *Germans into Nazis* (Cambridge, MA: Harvard University Press, 1998); Weinberg, *Foreign Policy of Hitler's Germany*, vol. 1: *Diplomatic Revolution*; Saul Friedländer, *Nazi Germany and the Jews*, vol. 1: *The Years of Persecution, 1933–1939* (New York: HarperCollins, 1997); Alexander Rossino, *Hitler Strikes Poland: Blitzkrieg, Ideology, and Atrocity* (Lawrence: University Press of Kansas, 2003); and many others.

21. Kershaw, *Hitler*, 1: xxviii; 2: xli.

22. Weikart, *From Darwin to Hitler*, ch. 7.

23. Weikart, *From Darwin to Hitler*. For an excellent discussion of Haeckel's views on evolutionary ethics, see Jürgen Sandmann, *Der Bruch mit der humanitären Tradition. Die Biologisierung der Ethik bei Ernst Haeckel und anderen Darwinisten seiner Zeit* (Stuttgart: Gustav Fischer Verlag, 1990). On Haeckel's social Darwinism, see Daniel Gasman, *The Scientific Origins of National Socialism: Social Darwinism in Ernst Haeckel and the German Monist League* (London: MacDonald, 1971).

24. Hamann, *Hitler's Vienna*, 84, 102, 202–203.

25. Wilfried Daim, *Der Mann, der Hitler die Ideen Gab: Jörg Lanz von Liebenfels*, 3rd ed. (Vienna: Ueberreuter, 1994).

26. For more on Lanz von Liebenfels' views, see Weikart, *From Darwin to Hitler*, 217–219.

27. Ludwig Woltmann, *Politische Anthropologie: Eine Untersuchung über den Einfluss der Deszendenztheorie auf die Lehre von der politischen Entwicklung der Völker* (Jena: Eugen Diederichs, 1903), 266.

28. Ludwig Woltmann and Hans K. E. Buhmann, "Naturwissenschaft und Politik," *Politisch-anthropologische Revue* 1 (1902): 1.

29. Walkenhorst, *Nation—Volk—Rasse*, 119–128, quote at 119.

30. Ernst Rüdin, review of Ludwig Woltmann, *Die Germanen und die Renaissance in Italien*, in *Archiv für Rassen- und Gesellschaftsbiologie* 1 (1904): 309; Rüdin, review of Ludwig Woltmann, *Politische Anthropologie*, in *Archiv für Rassen- und Gesellschaftsbiologie* 2 (1905): 609–19; Rüdin, review of Ludwig Woltmann, *Die Germanen in Frankreich*, in *Archiv für Rassen- und Gesellschaftsbiologie* 4 (1907): 234–238.

31. Eugen Fischer, "Sozialanthropologie," in *Handwörterbuch der Naturwissenschaften* (Jena: Gustav Fischer, 1912–1913), 9: 177; Niels Lösch, *Rasse als Konstrukt: Leben und Werk Eugen Fischers* (Frankfurt: Lang, 1997), 103.

32. Ludwig Schemann, *Die Rasse in den Geisteswissenschaften: Studien zur Geschichte des Rassengedankens*, vol. 3: *Die Rassenfragen im Schrifttum der Neuzeit*, 2nd ed. (Munich: J. F. Lehmann, 1943), xi, xvi, 229–241, 434; Ludwig Schemann, *Lebensfahrten eines Deutschen* (Leipzig: Erich Matthes, 1925), 295–297.

33. Ludwig Woltmann, *Die Germanen und die Renaissance in Italien* (Leipzig: Thüringische Verlagsanstalt, 1905), in Adolf Hitler's Personal Library, in Third Reich Collection, US Library of Congress.

34. Otto Reche, "Ludwig Woltmann," in Ludwig Woltmann, *Werke*, vol. 1: *Politische Anthropologie*, ed. Otto Reche (Leipzig: Justus Dörner, 1936), 7–8.

35. Rainer Hering, *Konstruierte Nation: Der Alldeutsche Verband 1890 bis 1939* (Hamburg: Christians, 2003), 482.

36. Fritz Lenz, *Die Rasse als Wertprinzip, Zur Erneuerung der Ethik* (Munich: J. F. Lehmann, 1933), 6–7. Hitler's library had the original 1917 version.
37. Rudolf von Sebottendorff, *Bevor Hitler kam: Urkundliches aus der Frühzeit der nationalsozialistischen Bewegung* (Munich: Deukula Verlag, 1933), 184–186.
38. NSDAP Rundschreiben, March 4, 1922 [signed by Hitler], in *Hitler: Sämtliche Aufzeichnungen*, 595.
39. *Fünfzig Jahre J. F. Lehmanns Verlag 1890–1940: Zur Erinnerung an das fünfzigjährige Bestehen am 1. September 1940* (Munich: J. F. Lehmanns Verlag, 1940), 81, 85, 171.
40. Paul Weindling, "The Medical Publisher Julius Friedrich Lehmann and the Racialising of German Medicine, 1890–1945," in *Die 'rechte Nation' und ihr Verleger. Politik und Popularisierung im J. F. Lehmanns Verlag, 1890–1979*, ed. Sigrid Stöckel (Berlin: Lehmanns, 2002), 169. See also Klaus-Dieter Thomann, "Dienst am Deutschtum—der medizinische Verlag J. F. Lehmanns und der Nationalsozialismus," in *Medizin im "Dritten Reich,"* ed. Johanna Bleker and Norbert Jachertz, 2nd ed. (Cologne: Deutscher Ärzte-Verlag, 1993), 54–69.
41. Some examples are Friedländer, *Nazi Germany and the Jews*, 1: 87; Christopher Browning with contributions by Jürgen Matthäus, *The Origins of the Final Solution: The Evolution of Nazi Jewish Policy, September 1939–March 1942* (Lincoln: University of Nebraska Press, 2004), 10; Peter Longerich, *The Unwritten Order: Hitler's Role in the Final Solution* (Stroud, UK: Tempus, 2003), 38; Robert S. Wistrich, *Hitler and the Holocaust* (New York: Modern Library, 2003), xii; David Welch, *The Third Reich: Politics and Propaganda*, 2nd ed. (London: Routledge, 2002), 94; Robert Edwin Herzstein, *The War that Hitler Won: The Most Infamous Propaganda Campaign in History* (New York: G. P. Putnam's Sons, 1978), 22; Lucy Dawidowicz, *The War against the Jews* (New York: Bantam Books, 1975), 5, 219; Daniel Goldhagen, *Hitler's Willing Executioners: Ordinary Germans and the Holocaust* (New York: Knopf, 1996), 9.
42. Michael Kellogg, *The Russian Roots of Nazism: White Émigrés and the Making of National Socialism, 1917–1945* (Cambridge: Cambridge University Press, 2005).
43. On the connection between anti-Semitism and social Darwinism, see Richard Weikart, "The Impact of Social Darwinism on Anti-Semitic Ideology in Germany and Austria, 1860–1945," in *Jewish Tradition and the Challenge of Evolution*, ed. Geoffrey Cantor and Mark Swetlitz (Chicago: University of Chicago Press, 2006), 93–115.

Chapter 1

1. Scholars debate the relationship of Nazism to conservatism and modernism. For differing perspectives, see Roger Griffin, *Modernism and Fascism: The Sense of a Beginning under Mussolini and Hitler* (New York: Palgrave Macmillan, 2007); Jeffrey Herf, *Reactionary Modernism: Technology, Culture, and Politics in Weimar and the Third Reich* (Cambridge: Cambridge University Press, 1984); Neil Gregor, *How to Read Hitler* (New York: Norton, 2005); Michael Burleigh and Wolfgang Wippermann, *The Racial State: Germany, 1933–1945* (Cambridge: Cambridge University Press, 1991); Zygmunt Bauman, *Modernity and the Holocaust* (Ithaca: Cornell University Press, 1989); Michael Prinz and Rainer Zitelmann, eds., *Nationalsozialismus und Modernisierung* (Darmstadt: Wissenschaftliche Buchgesellschaft, 1991); Richard Rubenstein, "Modernization and the Politics of Extermination," in *A Mosaic of Victims*, ed. Michael Berenbaum (New York: New York University Press, 1990), 3–19; Rainer Zitelmann, *Hitler: Selbstverständnis eines Revolutionärs.* (Hamburg: Berg, 1987); Norbert Frei "Wie modern war der Nationalsozialismus," *Geschichte und Gesellschaft* 19 (1993): 367–387.
2. Adolf Hitler to Adolf Gemlich, September 16, 1919, in *Hitler: Sämtliche Aufzeichnungen, 1905–1924*, ed. Eberhard Jäckel (Stuttgart: Deutsche Verlags-Anstalt, 1980), 90.

3. Adolf Hitler, speech to NSDAP-Jugendbund, "Pflicht, Treue, Gehorsam," December 3, 1922, in ibid., 752.

4. Adolf Hitler, *Adolf Hitler spricht: Ein Lexikon des Nationalsozialismus* (Leipzig: R. Kittler Verlag, 1934), 73.

5. Hitler, speech on October 14, 1933, in *Hitler: Speeches and Proclamations, 1932–1945*, ed. Max Domarus, 4 vols. (Wauconda, IL: Bolchazy-Carducci Publishers, 1990), 1: 369; see also Georg Usadel, *Zucht und Ordnung: Grundlagen einer nationalsozialistischen Ethik*, 3rd ed. (Hamburg: Hanseatische Verlagsanstalt, 1935).

6. Quoted in Michael Burleigh, *The Third Reich: A New History* (New York: Hill and Wang, 2000), 201.

7. Robert Gellately, *Backing Hitler: Consent and Coercion in Nazi Germany* (Oxford: Oxford University Press, 2001), ch. 3.

8. Hitler, Adolf. "Warum musste ein 8. November kommen?" *Deutschlands Erneuerung* 8 (April 1924): 207; emphasis in original.

9. Claudia Koonz, *The Nazi Conscience* (Cambridge: Belknap Press of Harvard University Press, 2003), 17–18.

10. Joachim Ribbentrop, *The Ribbentrop Memoirs* (London: Weidenfeld and Nicolson, 1954), 32.

11. *Die Tagebücher von Joseph Goebbels*, ed. Elke Fröhlich, part 1, vol. 7: *Juli 1939–März 1940* (Munich: K. G. Saur, 1987ff.), 150.

12. Hitler, orders to SA chief Lutz, June 30, 1934, in Domarus, *Hitler*, 1: 475–476.

13. Domarus, *Hitler*, 1: 246.

14. Hitler, *Mein Kampf*, trans. Ralph Manheim (Boston: Houghton Mifflin, 1943), 324.

15. Ibid., 661, 232, 305, 351–352.

16. Ibid., 384.

17. Ibid., 468.

18. Hitler, speech in Berlin, October 3, 1941, in Domarus, *Hitler*, 4: 2489.

19. Hitler, *Mein Kampf*, 178.

20. Hitler, speech on April 1, 1939, in Domarus, *Hitler*, 3: 1525.

21. Randy Bytwerk, *Bending Spines: The Propagandas of Nazi Germany and the German Democratic Republic* (East Lansing: Michigan State University Press, 2004), 44.

22. Hitler, *Mein Kampf*, 182.

23. Ibid., 459.

24. Hitler, speech on February 24, 1935, in Domarus, *Hitler*, 2: 642.

25. Hitler, speech to Reichstag, January 30, 1939, in ibid, 3: 1447.

26. "Information given to the Supreme Commander of the Army [Von Brauchitsch] by the Fuehrer on 25 March 1939," in *Nazi Conspiracy and Aggression* (Washington, DC, 1946), 7: 84.

27. "Hitler vor Vertretern der deutschen Presse am 10. November 1938 in München, " in Adolf Hitler, *"Es spricht der Führer": 7 exemplarische Hitler-Reden*, ed. Hildegard von Kotze and Helmut Krausnick (Gütersloh: Sigbert Mohn Verlag, 1966), 269–270.

28. "Hitler vor Kreisleitern auf der Ordensburg Volgelsang am 29. April 1937," in ibid., 167–168.

29. Ibid., 169–170.

30. Hitler, speech on November 23, 1939 to Oberbefehlshaber in Berlin, in Domarus, 3: 1886.

31. Hitler, in ibid., 3: 1705.

32. Hitler, speech to generals at Obersalzberg, August 22, 1939, in ibid., 3: 1668.

33. Hitler, *Mein Kampf*, 615; emphasis in original.

34. Dietrich Eckart, *Der Bolschewismus von Moses bis Lenin. Zwiegespräch zwischen Adolf Hitler und mir* (Munich: Hoheneichen Verlag, 1924), 33.

35. Claudia Koonz argues for this explicitly in *Nazi Conscience*, and Burleigh and Wippermann, *Racial State*, also show the centrality of race in Nazi social thought and policy.

36. Hitler's discussion at NSDAP meeting, Munich, February 9, 1922, in *Hitler: Sämtliche Aufzeichnungen*, 569.

37. Karl Alexander von Müller, *Im Wandel einer Welt: Erinnerungen 1919–1932* (Munich: Süddeutscher Verlag, 1966), 148.

38. Hitler, "Politik und Rasse. Warum sind wir Antisemiten?" in *Hitler: Sämtliche Aufzeichnungen*, 909.

39. Hitler, "Die Proklamation des Führers: Vollstrecker des Willens der Nation" (Abschrift von der *Fränkische Tageszeitung*, September 6, 1934), in Hoover Institution, NSDAP Hauptarchiv, Reel 2, Folder 51, 4, 9.

Chapter 2

1. Othmar Plöckinger, *Geschichte eines Buches: Adolf Hitlers "Mein Kampf" 1922–1945* (Munich: R. Oldernbourg, 2006), 38–39, 86–88.

2. Hitler, *Mein Kampf*, trans. Ralph Manheim (Boston: Houghton Mifflin, 1943), 244–245.

3. Hitler, "Zukunft oder Untergang," March 6, 1927, 2, in Hoover Institution, NSDAP Hauptarchiv, Reel 2, Folder 59.

4. Charles Darwin, *The Origin of Species* (London: Penguin, 1968), 459.

5. Charles Darwin, *The Descent of Man*, 1st edition, 2 vols. in 1 (Reprint. Princeton: Princeton University Press, 1981), 1: 180.

6. Ibid., 2: 403.

7. Richard Weikart, "Was Darwin or Spencer the Father of Laissez-Faire Social Darwinism?" *Journal of Economic Behavior and Organization* (special issue on social Darwinism) (forthcoming); Weikart, "Laissez-Faire Social Darwinism and Individualist Competition in Darwin and Huxley," *The European Legacy* 3 (1998): 17–30; Weikart, "A Recently Discovered Darwin Letter on Social Darwinism," *Isis* 86 (1995): 609–611.

8. Friedrich Hellwald, *Culturgeschichte in ihrer natürlichen Entwicklung bis zur Gegenwart* (Augsburg: Lampart, 1875), 27; "Der Kampf ums Dasein im Menschen- und Völkerleben," *Das Ausland* 45 (1872): 105.

9. Peter Walkenhorst, *Nation—Volk—Rasse: Radikaler Nationalismus im Deutschen Kaiserreich 1890–1914* (Göttingen: Vandenhoeck and Ruprecht, 2007), 121–122.

10. Because of this, I will render "Entwicklung" as "evolution" in those passages of *Mein Kampf* and elsewhere, as long as the context makes clear that this is more accurate.

11. Hitler, "Der Nationalsozialismus als Weltanschauung, der Marxismus ein Wahnsinn!" (April 2, 1927), in *Hitler: Reden, Schriften, Anordnungen, Februar 1925 bis Januar 1933* (Munich: K. G. Saur, 1992–1995), 1: 228–229.

12. *Hitler's Second Book: The Unpublished Sequel to* Mein Kampf, ed. Gerhard Weinberg (New York: Enigma Books, 2003), 6–7. I have translated Lebenskampf more literally as struggle for life here, rather than struggle for survival.

13. Hitler, *Mein Kampf*, 285.

14. Hitler, "Wesen und Ziele des Nationalsozialismus," July 3, 1927, in *Hitler: Reden, Schriften, Anordnungen*, 2: 406.

15. Hitler, "Was ist Nationalsozialismus?" August 6, 1927, in *Hitler: Reden, Schriften, Anordnungen*, 2: 439, 462–463.

16. Hitler, "War der Zweite Weltkrieg für Deutschland vermeidbar?" May 30, 1942, in *Hitlers Tischgespräche im Führerhauptquartier*, ed. Henry Picker (Frankfurt: Ullstein, 1989), 491.

17. Ibid., 492.

18. Hitler, speech of November 8, 1943, in *Hitler: Speeches and Proclamations, 1932–1945*, ed. Max Domarus, 4 vols. (Wauconda, IL: Bolchazy-Carducci Publishers, 1990), 3: 2841.

19. Hitler, speech of November 12, 1944, in Domarus, 4: 2964.

20. Hitler, proclamation of February 24, 1945, in Domarus, 4: 3015.

21. Richard Steigmann-Gall, *The Holy Reich: Nazi Conceptions of Christianity, 1919–1945* (Cambridge: Cambridge University Press, 2003), ch. 1.

22. Hitler, speech of February 15, 1942, in *"Es spricht der Führer": 7 exemplarische Hitler-Reden*, ed. Hildegard von Kotze and Helmut Krausnick (Gütersloh: Sigbert Mohn Verlag, 1966), 307–308; Hitler made the same points in his May 3, 1940 speech in Domarus, 3: 1979.

23. Otto Dietrich, *The Hitler I Knew*, trans. Richard and Clara Winston (London: Methuen, 1957), 19.

24. Traudl Junge, *Bis zur letzten Stunde: Hitlers Sekretärin erzählt ihr Leben* (Munich: Claassen Verlag, 2002), 122.

25. Otto Wagener, *Hitler—Memoirs of a Confidant*, ed. Henry Ashby Turner, trans. Ruth Hein (New Haven: Yale University Press, 1985), 114, 40.

26. Nicolaus von Below, *Als Hitlers Adjutant 1937–1945* (Mainz: v. Hase and Koehler Verlag, 1980), 360.

27. Hans Frank, *Im Angesicht des Galgens: Deutung Hitlers und seiner Zeit auf Grund eigener Erlebnisse und Erkenntnisse* (Munich-Gräfelfing: Friedrich Alfred Beck Verlag, 1953), 238.

28. Hitler, "Was ist Nationalsozialismus?" August 6, 1927, in *Hitler: Reden, Schriften, Anordnungen*, 2: 439–440.

29. Hitler, "Tageskampf oder Schicksalskampf," March 3, 1928, in *Hitler: Reden, Schriften, Anordnungen*, 2: 726.

30. Ibid.

31. Hitler, "Weltjude und Weltbörse, die Urschuldigen am Weltkriege," April 13, 1923, in *Hitler: Sämtliche Aufzeichnungen, 1905–1924*, ed. Eberhard Jäckel (Stuttgart: Deutsche Verlags-Anstalt, 1980), 887.

32. Hitler, "Was ist Nationalsozialismus?" August 6, 1927, in *Hitler: Reden, Schriften, Anordnungen*, 2: 439, 462–463.

33. Hitler, in Domarus, 1: 89–90.

34. Hitler, *Mein Kampf*, 343.

35. *Hitler's Second Book*, 17.

36. Hitler, *Mein Kampf*, 287–288.

37. See Richard Weikart, *From Darwin to Hitler: Evolutionary Ethics, Eugenics, and Racism in Germany* (New York: Palgrave Macmillan, 2004), chs. 1–2.

38. The misconception that Hitler rejected evolution is widespread on the Internet. Mike Hawkins, *Social Darwinism in European and American Thought, 1860–1945: Nature as Model and Nature as Threat* (Cambridge: Cambridge University Press, 1997), 283, calls Hitler ambivalent about human descent from apes, because of this passage, rightly noting that in other passages Hitler accepted the idea. However, Maser quotes this passage, claiming that Hitler did not believe in human evolution at all (though Maser also calls Darwin one of Hitler's "teachers," since Hitler embraced the idea of the struggle for existence), Werner Maser, *Hitlers Briefe und Notizen. Sein Weltbild in handschriftlichen Dokumenten* (Düsseldorf: Econ Verlag, 1973), 301.

39. *Hitlers Tischgespräche*, 93–94.

40. Gerhard Weinberg, *The Foreign Policy of Hitler's Germany*, vol. 1: *Diplomatic Revolution in Europe, 1933–1936* (Chicago: University of Chicago Press, 1970), 4.

41. *The Speeches of Adolf Hitler, April 1922–August 1939*, ed. Norman H. Baynes, 2 vols. (Oxford: Oxford University Press, 1942), 1: 464.

42. Hitler, "Der Nationalsozialismus als Weltanschauung, der Marxismus ein Wahnsinn!" April 2, 1927, in *Hitler: Reden, Schriften, Anordnungen*, 2: 228–229.

43. *Die Tagebücher von Joseph Goebbels*, ed. Elke Fröhlich (Munich: K. G. Saur, 1987ff.), part 1, vol. 7: *Juli 1939–März 1940*, 250.

44. Hitler, "Was ist Nationalsozialismus?" August 6, 1927, in *Hitler: Reden, Schriften, Anordnungen*, 2: 442.
45. *Hitlers Tischgespräche*, 73–74.
46. Ibid., 76.
47. Ibid., 75.
48. Junge, *Bis zur letzten Stunde*, 122.
49. *Hitler's Second Book*, 9.
50. Ibid., 19.
51. Hitler, *Mein Kampf*, 287.
52. Hitler, "Ansprache des Führeres vor Generalen und Offiziers am 22.6.1944 im Platterhof," 2, in Hoover Institution, NSDAP Hauptarchiv, Reel 2, Folder 51.
53. Ibid., 3–4.
54. Hitler, *Mein Kampf*, 402, 405.
55. J. Lanz-Liebenfels, "Moses als Darwinist, eine Einführung in die anthropologische Religion," *Ostara* 2nd ed., no. 46 (1917), 16.
56. Hitler, *Mein Kampf*, 444.
57. Ibid., 287–288.
58. Ibid., 177–178.
59. Hitler, "Warum sind wir Antisemiten?" August 13, 1920, in *Hitler: Sämtliche Aufzeichnungen*, 184–185.
60. Ibid., 185–186.
61. Ibid., 185.

Chapter 3

1. Hans F. K. Günther, *Mein Eindruck von Adolf Hitler* (Pähl, F. von Bebenburg, 1969), 18–21.
2. Hitler to unknown person, February 2, 1930, in *Hitler: Reden, Schriften, Anordnungen, Februar 1925 bis Januar 1933*, vol. 3, part 3: *Januar 1930–September 1930* (Munich: K. G. Saur, 1992–2003), 61–62.
3. Fritsch's inscription to Hitler in Theodor Fritsch, *Mein Streit mit dem Hause Warburg: Eine Episode aus dem Kampfe gegen das Weltkapital* (Leipzig: Hammer-Verlag, 1925), Adolf Hitler's Personal Library, Third Reich Collection, United States Library of Congress.
4. Hitler to Theodor Fritsch, November 28, 1930, in *Hitler: Reden, Schriften, Anordnungen*, vol. 4, part 1: 133.
5. Hitler "10 Jahre Kampf," *Illustrierter Beobachter*, August 3, 1929, in *Hitler: Reden, Schriften, Anordnungen*, vol. 3, part 2: 341–342.
6. Hitler, *Mein Kampf*, trans. Ralph Manheim (Boston: Houghton Mifflin, 1943), 383.
7. Benoit Massin, "From Virchow to Fischer: Physical Anthropology and 'Modern Race Theories' in Wilhelmine Germany," in *Volksgeist as Method and Ethic*, ed. George W. Stocking (Madison: University of Wisconsin Press, 1996), 79–154; Barry W. Butcher, "Darwinism, Social Darwinism and the Australian Aborigines: A Reevaluation," in *Darwin's Laboratory: Evolutionary Theory and Natural History in the Pacific*, ed. Roy MacLeod and Philip F. Rehbock (Honolulu: University of Hawaii Press, 1994), 371–394.
8. For a more detailed discussion, see Richard Weikart, *From Darwin to Hitler: Evolutionary Ethics, Eugenics, and Racism in Germany* (New York: Palgrave Macmillan, 2004), chs. 6, 10; or Weikart, "Progress through Racial Extermination: Social Darwinism, Eugenics, and Pacifism in Germany, 1860–1918," *German Studies Review* 26 (2003): 273–294.
9. Ernst Haeckel, *Die Lebenswunder: Gemeinverständliche Studien über Biologische Philosophie* (Stuttgart: Alfred Kröner, 1904), 450; Benoit Massin, "Rasse und Vererbung als Beruf: Die

Hauptforschungsrichtungen am Kaiser-Wilhlem-Institut für Anthropologie, menschliche Erblehre und Eugenik im Nationalsozialismus," in *Rassenforschung an Kaiser-Wilhelm-Instituten vor und nach 1933*, ed. Hans-Walter Schmuhl (Göttingen: Wallstein Verlag, 2003), 193.

10. Massin, "From Virchow to Fischer," 99.

11. Charles Darwin, *The Descent of Man*, 1st edition, 2 vols. in 1 (Reprint. Princeton: Princeton University Press, 1981), 1: 201.

12. Michael Burleigh and Wolfgang Wippermann, *The Racial State: Germany, 1933–1945* (Cambridge: Cambridge University Press, 1991), is essentially a book-length treatment of this theme.

13. *Hitler's Second Book: The Unpublished Sequel to* Mein Kampf, ed. Gerhard Weinberg (New York: Enigma Books, 2003), 112.

14. Othmar Plöckinger, *Geschichte eines Buches: Adolf Hitlers "Mein Kampf" 1922–1945* (Munich: R. Oldernbourg, 2006), 12–13, 412–414.

15. Hitler, *Mein Kampf*, 289.

16. "Hitler vor Vertretern der deutschen Presse am 10. November 1938 in München," in Adolf Hitler, *"Es spricht der Führer": 7 exemplarische Hitler-Redeni*, ed. Hildegard von Kotze and Helmut Krausnick (Gütersloh: Sigbert Mohn Verlag, 1966), 285.

17. Hitler, speech on May 1, 1935, in *Hitler: Speeches and Proclamations, 1932–1945*, ed. Max Domarus, 4 vols. (Wauconda, IL: Bolchazy-Carducci Publishers, 1990), 2: 664–665.

18. Many scholars acknowledge this, including Eberhard Jäckel, *Hitler's Weltanschauung: A Blueprint for Power*, trans. Herbert Arnold (Middleton, CT: Wesleyan University Press, 1972), ch. 5.

19. Hitler, "Entwicklung unserer Bewegung," January 7, 1922, in *Hitler: Sämtliche Aufzeichnungen, 1905–1924*, ed. Eberhard Jäckel (Stuttgart: Deutsche Verlags-Anstalt, 1980), 541.

20. Hitler, *Mein Kampf*, 388–389; see also 312.

21. *Hitler's Second Book*, 49.

22. *Hitlers Zweites Buch: Ein Dokument aus dem Jahr 1928*, ed. Gerhard L. Weinberg (Stuttgart: Deutsche Verlags-Anstalt, 1961), 65.

23. Hitler, "Weltanschauung und Kommunalpolitik," November 29, 1929, in *Hitler: Reden, Schriften, Anordnungen, Februar 1925 bis Januar 1933* (Munich: K. G. Saur, 1992–1995), 3: 485–486.

24. Hitler, "Die soziale Sendung des Nationalsozialismus," December 16, 1925, in *Hitler: Reden, Schriften, Anordnungen*, vol. 1: 258.

25. Hitler, speech on August 7, 1920, in *Hitler: Sämtliche Aufzeichnungen*, 177.

26. Hitler, "Die 'Hetzer' der Wahrheit," April 12, 1922, in ibid., 624.

27. Quoted in J. Noakes and G. Pridham, *Nazism 1919–1945: A Documentary Reader*, vol. 2: *State, Economy and Society 1933–1939* (Exeter: University of Exeter Press, 2000), 343.

28. See Hitler, *Mein Kampf*, 3, for his Pan-German views.

29. Phillip T. Rutherford, *Prelude to the Final Solution: The Nazi Program for Deporting Ethnic Poles, 1939–1941* (Lawrence: University Press of Kansas, 2007), 8–9.

30. Hitler, decree on July 28, 1942, in Domarus, 4: 2657.

31. Cornelia Essner, *Die "Nürnberger Gesetze" oder Die Verwaltung des Rassenwahns 1933–1945* (Paderborn: Ferdinand Schöningh, 2002), 173.

32. "Ansprache Hitlers vor Generalen und Offizieren am 26. Mai 1944 im Platterhof," in Hans-Heinrich Wilhelm, "Hitlers Ansprache vor Generalen und Offizieren am 26. Mai 1944," *Militärgeschichtliche Mitteilungen* 2 (1976): 148–149.

33. *Rassenpolitik* (Berlin: Der Reichsführer SS, SS-Hauptamt, n.d.).

34. Essner, *"Nürnberger Gesetze,"* 173; Christopher Hutton, *Race and the Third Reich: Linguistics, Racial Anthropology and Genetics in the Dialectic of Volk* (Cambridge, UK: Polity, 2005), 3, 23, 34–36, 60–61, 80.

35. *Hitlers Tischgespräche im Führerhauptquartier*, ed. Henry Picker (Frankfurt: Ullstein, 1989), 70.

36. Essner, *"Nürnberger Gesetze,"* 280–281.

37. John Connelly, "Nazis and Slavs: From Racial Theory to Racist Practice," *Central European History* 32 (1999): 14.

38. Vojtech Mastny, *The Czechs under Nazi Rule: The Failure of National Resistance, 1939–1942* (New York: Columbia University Press, 1971), 128.

39. Chad Bryant, *Prague in Black: Nazi Rule and Czech Nationalism* (Cambridge, MA: Harvard University Press, 2007), 113, 125–126, 156–161.

40. Ibid., 116.

41. Diemut Majer, *"Non-Germans" under the Third Reich: The Nazi Judicial and Administrative System in Germany and Occupied Eastern Europe with Special Regard to Occupied Polande, 1939–1945*, trans. Peter Thomas Hill et al. (Baltimore: Johns Hopkins University Press, 2003), 238–239.

42. Hitler, decree of March 12, 1942, in *Führer-Erlasse" 1939–1945*, ed. Martin Moll (Stuttgart: Franz Steiner Verlag, 1997), 240.

43. Egbert Klautke, "German 'Race Psychology' and Its Implementation in Central Europe: Egon von Eickstedt and Rudolf Hippius," in *Blood and Homeland: Eugenics and Racial Nationalism in Central and Southeast Europe 1900–1940*, ed. Marius Turda and Paul Weindling (Budapest: Central European University Press, 2007), 23–40; Gretchen E. Schafft, *From Racism to Genocide: Anthropology in the Third Reich* (Urbana: University of Illinois Press, 2004), 132.

44. Hitler, speech at the Industrieklub in Düsseldorf, January 27, 1932, in Domarus, 1: 91.

45. Hitler, *Mein Kampf*, 296.

46. Ibid., 430. See also Hitler, "Weltanschauung und Kommunalpolitik," November 29, 1929, in *Hitler: Reden, Schriften, Anordnungen*, vol. 3, part 2: 486–487.

47. Hitler, *Mein Kampf*, 403.

48. Baldur von Schirach, *Ich glaubte an Hitler* (Hamburg: Mosaik Verlag, 1967), 217; see also David Clay Large, "'Darktown Parade': African Americans in the Berlin Olympics of 1936," *Historically Speaking* 9, 2 (November/December 2007): 7.

49. Hitler, *Mein Kampf*, 659.

50. Ibid., 290–291.

51. Richard Overy, *Interrogations: The Nazi Elite in Allied Hands, 1945* (New York: Viking, 2001), 332.

52. Hitler, *Mein Kampf*, 158.

53. Hitler, speech to Reichstag, January 30, 1939, in Domarus, 3: 1064.

54. Hitler, *Mein Kampf*, 146.

55. Ibid., 524, 654–655.

56. *Hitler's Second Book*, 151.

57. Joseph Stalin, "Report to the Seventeenth Party Congress," January 26, 1934, in J. V. Stalin, *Works* (Moscow: Foreign Languages Publishing House, 1955), 13: 302.

58. Hitler, *Mein Kampf*, 390.

59. *Hitler's Second Book*, 49, 53.

60. Hitler, speech in Danzig, September 19, 1939, in Domarus, 3: 1802.

61. Connelly, "Nazis and Slavs"; Connelly is incorrect, however, in his claim that Hitler did not consider Poles inferior until 1939.

62. Martyn Housden, *Hans Frank: Lebensraum and the Holocaust* (Houndmills: Palgrave Macmillan, 2003), 96.

63. Joachim C. Fest, *The Face of the Third Reich: Portraits of the Nazi Leadership*, trans. Michael Bullock (New York: Pantheon, 1970), 115.

64. *Hitlers Tischgespräche*, 215–216.

65. Majer, *"Non-Germans" under the Third Reich*, 210.

66. Joseph Goebbels, December 29, 1939, *Die Tagebücher von Joseph Goebbels*, ed. Elke Fröhlich (Munich: K. G. Saur, 1987ff.), part 1, vol. 7: 250.

67. Franz Halder, *Kriegstagebuch*, ed. Hans-Adolf Jacobsen, vol. 2: *Von der geplanten Landung in England bis zum Beginn des Ostfeldzuges (1.7.1940–21.6.1941)* (Stuttgart: W. Kohlhammer Verlag, 1963), 214.

68. *Hitlers Tischgespräche*, 421.

69. Hitler, *Mein Kampf*, 290–296.

70. Erwin Baur, "Der Untergang der Kulturvölker im Lichte der Biologie," *Deutschlands Erneuerung* 6 (1922): 257–268.

71. Hitler, *Mein Kampf*, 290–296; see also Hitler, "Der Nationalsozialismus als Weltanschauung, der Marxismus ein Wahnsinn!" April 2, 1927, in *Hitler: Reden, Schriften, Anordnungen*, vol. 2, Part 1: 228–229.

72. Richard Weikart, "The Impact of Social Darwinism on Anti-Semitic Ideology in Germany and Austria, 1860–1945," in *Jewish Tradition and the Challenge of Evolution*, ed. Geoffrey Cantor and Mark Swetlitz (Chicago: University of Chicago Press, 2006), 93–115.

73. Hitler, *Mein Kampf*, 300–325.

74. Hitler, speech on June 12, 1925, in *Hitler: Reden, Schriften, Anordnungen*, vol. 1: 92.

75. Hitler, *Mein Kampf*, 339.

76. Hitler, speech in Landshut, May 5, 1922, in *Hitler: Sämtliche Aufzeichnungen*, 638; see also Hitler's notes for a speech in Munich, March 1, 1922, in *Hitler: Sämtliche Aufzeichnungen*, 588.

77. Hitler, *Mein Kampf*, 295.

78. Ibid., 393.

79. Ibid., 289.

80. Hitler, "Was ist Nationalsozialismus?" August 6, 1927, in *Hitler: Reden, Schriften, Anordnungen*, vol. 2: 440.

81. Hitler, "Warum sind wir Antisemiten?" August 13, 1920, in *Hitler: Sämtliche Aufzeichnungen*, 186.

82. Hitler, speech on December 11, 1941, in Domarus, 4: 2534.

83. Hitler, speech to Düsseldorf Industrieklub on January 27, 1932, in Domarus, 1: 91.

84. *Hitlers Tischgespräche*, 101; see also 85, 93–94.

85. Otto Dietrich, *The Hitler I Knew*, trans. Richard and Clara Winston (London: Methuen, 1957) 191; Hans Frank, *Im Angesicht des Galgens: Deutung Hitlers und seiner Zeit auf Grund eigener Erlebnisse und Erkenntnisse* (Munich-Gräfelfing: Friedrich Alfred Beck Verlag, 1953), 313.

86. Alfred Rosenberg, *Letzte Aufzeichnungen: Nürnberg 1945/46*, 2nd ed. (Uelzen: Jomsburg-Verlag, 1996), 96.

87. Goebbels, April 8, 1941, in *Tagebücher*, part 1, vol. 9: 234.

88. Frederic Spotts, *Hitler and the Power of Aesthetics* (Woodstock: Overlook Press, 2003), 15, 399.

89. Hitler, speech on September 7, 1937, in Domarus, 2: 927.

90. Hitler, speech on July 19, 1937, in Noakes and Pridham, *Nazism*, 2: 205–206.

91. Otto Wagener, *Hitler—Memoirs of a Confidant*, ed. Henry Ashby Turner, trans. Ruth Hein (New Haven: Yale University Press, 1985), 168.

92. Hitler, "Der Nationalsozialismus als Weltanschauung, der Marxismus ein Wahnsinn!" April 2, 1927, in *Hitler: Reden, Schriften, Anordnungen*, 2: 228–229. Hitler made the same point in his *Second Book*, 12.

93. Hitler, speech on February 15, 1942, in *"Es spricht der Führer,"* 317–318.

94. Hitler, *Mein Kampf*, 393–395.

95. Ibid., 612.

96. *Hitler's Second Book*, 32–34.

97. Hitler, *Mein Kampf*, 177. I have translated "um ihre Existenz…kämpfen" more literally as "struggle for their existence."

98. Hitler, secret speech on November 23, 1937, in *Hitlers Tischgespräche*, 487.

99. Hitler, "Warum sind wir Antisemiten?" August 13, 1920, in *Hitler: Sämtliche Aufzeichnungen*, 202.

100. Hitler, "Weltanschauung und Kommunalpolitik," November 29, 1929, in *Hitler: Reden, Schriften, Anordnungen*, vol. 3, part 2: 485–486; emphasis in original.

101. Hitler, speech at the Industrieklub in Düsseldorf, January 27, 1932, in Domarus, 1: 95–96.

102. *Hitler's Second Book*, 17.

103. Hitler, speech on November 23, 1939, in Domarus, 3: 1885–1886; this is my translation of the German original.

Chapter 4

1. Hitler, *Mein Kampf*, trans. Ralph Manheim (Boston: Houghton Mifflin, 1943), 65; in the second and all subsequent editions, Hitler replaced "thousands of years ago" with "millions of years ago."

2. Ernst Haeckel, *Freie Wissenschaft und freie Lehre* (Stuttgart: E. Schweizerbart's che Verlagshandlung, 1878), 72–75; Haeckel, "Die Wissenschaft und der Umsturz," *Die Zukunft* 10 (1895): 205–206.

3. Willibald Hentschel, *Varuna: Das Gesetz des aufsteigenden und sinkenden Lebens in der Geschichte* (Leipzig: Theodor Fritsch, 1907), 14–15.

4. Theodor Fritsch to Ludwig Schemann, February 11, 1908, in Ludwig Schemann papers, IV B 1/2, University of Freiburg Library Archives; Dieter Löwenberg, "Willibald Hentschel (1858–1947): Seine Pläne zur Menschenzüchtung, sein Biologismus und Antisemitismus" (dissertation, University of Mainz, 1978), 46.

5. Houston Stewart Chamberlain, *Die Grundlagen des neunzehnten Jahrhunderts*, 2 vols. (Munich: F. Bruckmann, 1899), 1: 265–266, 278, 531; 2: 717.

6. Jeffrey Herf, *The Jewish Enemy: Nazi Propaganda during World War II and the Holocaust* (Cambridge, MA: Belknap Press of Harvard University Press), 2006.

7. Hitler, "Zwei Fronten in Deutschland" (January 18, 1923), in *Hitler: Sämtliche Aufzeichnungen, 1905–1924*, ed. Eberhard Jäckel (Stuttgart: Deutsche Verlags-Anstalt, 1980), 796.

8. Hitler, "Die 'Hetzer' der Wahrheit" (April 12, 1922), in *Adolf Hitlers Reden*, ed. Ernst Boepple (Munich: Deutscher Volksverlag Dr. E. Boepple, 1934), 17.

9. Hitler, *Mein Kampf*, 661.

10. Saul Friedländer, *Nazi Germany and the Jews*, vol. 1: *The Years of Persecution, 1933–1939* (New York: HarperCollins, 1997).

11. *Hitler's Second Book: The Unpublished Sequel to* Mein Kampf, ed. Gerhard L. Weinberg (New York: Enigma Books, 2003), 230.

12. Herf, *Jewish Enemy*, 151.

13. Richard F. Wetzell, *Inventing the Criminal: A History of German Criminology, 1880–1945* (Chapel Hill: University of North Carolina Press, 2000), esp. ch. 7; Georg Lilienthal, "Die jüdischen 'Rassenmerkmale': Zur Geschichte der Anthropologie der Juden," *Medizinhistorisches Journal* 28, 2/3 (1993): 173–198; on biological determinism of behavior see also the many works on eugenics in Germany.

14. Michael Berkowitz, *The Crime of My Very Existence: Nazism and the Myth of Jewish Criminality* (Berkeley: University of California Press, 2007), 44; see also Alan Steinweis, *Studying the Jew: Scholarly Antisemitism in Nazi Germany* (Cambridge, MA: Harvard University Press, 2006), 138–140.

15. Hitler, *Vortrag Adolf Hitlers von westdeutscher Wirtschaftlern im Industrie-Klub zu Dusseldorf am 27. Januar 1932*, 1st ed. (Munich: Frz. Eher Nachf., n.d. [probably 1932]), 7.

16. Hitler, *Mein Kampf*, 297.

17. Ibid., 299.

18. A couple of good discussions (among many) are Claudia Koonz, *The Nazi Conscience* (Cambridge, MA: Harvard University Press, 2003); Geoffrey Cocks, "Sick Heil: Self and Illness in Nazi Germany," *Osiris* 22 (2007): 93–115.

19. Hitler, May 3, 1940 speech, in *Hitler: Speeches and Proclamations, 1932–1945*, ed. Max Domarus, 4 vols. (Wauconda, IL: Bolchazy-Carducci Publishers, 1990), 3: 1986.

20. In Richard Weikart, *From Darwin to Hitler: Evolutionary Ethics, Eugenics, and Racism in Germany* (New York: Palgrave Macmillan, 2004), 81–83, I show that many evolutionists before Hitler made similar arguments.

21. Hitler monolog, December 1, 1941, in *Hitlers Tischgespräche im Führerhauptquartier*, ed. Henry Picker (Frankfurt: Ullstein, 1989), 79.

22. Hitler, *Mein Kampf*, 104, 164–165.

23. Ibid., 266.

24. Hitler, "Ein Kampf um Deutschlands Freiheit," February 5, in *Hitler: Reden, Schriften, Anordnungen, Februar 1925 bis Januar 1933*, 6 vols. (Munich: K. G. Saur, 1992–2003), vol. 2, part 2: 666.

25. Hitler, speech on October 1, 1933, in Domarus, 1: 363.

26. Hitler, *Mein Kampf*, 255.

27. Ibid., 254.

28. "Program of the National Socialist German Worker's Party," at www.yale.edu/lawweb/avalon/imt/nsdappro.htm, accessed September 28, 2001.

29. Report of Hitler's speech "Parteipolitik und Judenfrage" (December 8, 1920), in *Hitler: Sämtliche Aufzeichnungen*, 276.

30. Hitler, "Warum sind wir Antisemiten?" (August 13, 1920), in *Hitler: Sämtliche Aufzeichnungen*, 190.

31. Hitler, *Mein Kampf*, 416.

32. Ibid., 280.

33. Ibid., 324.

34. Ibid., 57.

35. Hitler, interview with George Sylvester Viereck, published in *The American Monthly*, October 1923, in *Hitler: Sämtliche Aufzeichnungen*, 1025.

36. Hitler, January 30, 1934 speech, in Domarus, 1: 354.

37. See Wetzell, *Inventing the Criminal*, esp. ch. 7, for a good discussion of the Nazi's biological determinism and crime.

38. Hitler, *Mein Kampf*, 301–302.

39. Hitler, speech on September 13, 1937, in Domarus, 2: 939.

40. Hitler, *Mein Kampf*, 59.

41. Report on Hitler's speech, "Das deutsche Weib und der Jude" (December 16, 1921), in *Hitler: Sämtliche Aufzeichnungen*, 531.

42. Hitler, *Mein Kampf*, 325; see also "Das deutsche Weib und der Jude," in *Hitler: Sämtliche Aufzeichnungen*, 531.

43. Hitler, "Die Wahrhaftigkeit als Grundlage des politischen Handelns" (June 22, 1922), in ibid., 645.

44. Hitler, "Die 'Hetzer' der Wahrheit" (April 12, 1922), in *Adolf Hitlers Reden*, 18–19.

45. Hitler, "Warum sind wir Antisemiten?" (August 13, 1920), in *Hitler: Sämtliche Aufzeichnungen*, 190.

46. Hitler, *Mein Kampf*, 308–327.

47. Hitler's speech on May 29, 1923, in *Hitler: Sämtliche Aufzeichnungen*, 931.

48. Hitler, *Mein Kampf*, 305; see also Hitler, "Was ist Nationalsozialismus?" (August 6, 1927), in *Hitler: Reden, Schriften, Anordnungen*, vol. 2, part 2: 462–463.

49. Hitler, *Mein Kampf*, 151.

50. Ibid., 153.

Chapter 5

1. Hitler, speech of October 5, 1937, in *Hitler: Speeches and Proclamations, 1932–1945*, ed. Max Domarus, 4 vols. (Wauconda, IL: Bolchazy-Carducci Publishers, 1990), 2: 954.

2. Hitler, speech of October 5, 1937, in ibid., 2: 954.

3. Hitler, speech of October 10, 1939, in ibid., 3: 1853–1854.

4. Hitler, speech of September 4, 1940, in ibid., 3: 2090.

5. Hitler, speech of September 13, 1933, in ibid., 1: 357–359.

6. Peter Walkenhorst, *Nation—Volk—Rasse: Radikaler Nationalismus im Deutschen Kaiserreich 1890–1914* (Göttingen: Vandenhoeck and Ruprecht, 2007), 102–103, 119–128.

7. Domarus, 1: 448.

8. Ibid., 2: 887.

9. Hitler, speech of October 5, 1937, in ibid., 2: 954–956.

10. Hitler, speech of October 5, 1938, in ibid., 2: 1219.

11. Hitler, speech of October 10, 1939 and September 4, 1940, in ibid., 3: 1851, 2087–2088.

12. Hitler, speech of October 8, 1935, in ibid., 2: 716–717.

13. Peter Fritzsche, *Germans into Nazis* (Cambridge, MA: Harvard University Press, 1998), 198–208.

14. Hitler, speech on February 10, 1933, in Domarus, 1: 247.

15. Hitler, speech on July 15, 1932, in ibid., 1: 145.

16. Hitler, speech on May 1, 1939, in ibid., 3: 1601.

17. Hitler, speech on October 9, 1938, in ibid., 2: 1222.

18. One example among many is "Die 'Hetzer' der Wahrheit" (April 12, 1922), in *Adolf Hitlers Reden*, ed. Ernst Boepple (Munich: Deutscher Volksverlag Dr. E. Boepple, 1934), 18–19.

19. See Hitler, recorded speech of July 15, 1932, in Domarus, 1: 144.

20. Hitler, "Warum sind wir Antisemiten?" (August 13, 1920), in *Hitler: Sämtliche Aufzeichnungen, 1905–1924*, ed. Eberhard Jäckel (Stuttgart: Deutsche Verlags-Anstalt, 1980), 200.

21. Hitler, "Free State and Slavery" (July 28, 1922), in *The Speeches of Adolf Hitler, April 1922–August 1939*, ed. Norman H. Baynes, 2 vols. (Oxford: Oxford University Press, 1942), 1: 26.

22. See, for example, "Was ist Nationalsozialismus?" (August 6, 1927), in *Hitler: Reden, Schriften, Anordnungen, Februar 1925 bis Januar 1933*, 6 vols. (Munich: K. G. Saur, 1992–2003), vol. 2, part 2: 440–442.

23. Hitler, "Rede vor dem Industrie-Club in Düsseldorf" (January 26, 1932), in *Hitler: Reden, Schriften, Anordnungen*, vol. 6, part 3: 77–80; English translation in Domarus, 1: 90–93.

24. Ibid.

25. *The Speeches of Adolf Hitler*, 1: 467–468.

26. Otto Wagener, *Hitler—Memoirs of a Confidant*, ed. Henry Ashby Turner, trans. Ruth Hein (New Haven: Yale University Press, 1985), 40–41.

27. Charles Darwin, *The Descent of Man*, 1st edition, 2 vols. in 1 (Reprint, Princeton: Princeton University Press, 1981), 1: 170–171, 180; 2: 403; see also Richard Weikart, "Laissez-Faire Social Darwinism and Individualist Competition in Darwin and Huxley," *The European Legacy* 3 (1998): 17–30; and "Was Darwin or Spencer the Father of Laissez-Faire Social Darwinism?" *Journal of Economic Behavior and Organization* (forthcoming).

28. Richard Weikart, "A Recently Discovered Darwin Letter on Social Darwinism," *Isis* 86 (1995): 609–611.

29. Ernst Haeckel, *Natürliche Schöpfungsgeschichte* (Berlin: Georg Reimer, 1868), 128–129, 226, 218–219; Haeckel, *Über Arbeitstheilung in Natur- und Menschenleben* (Berlin: C. G. Lüderitz'sche Verlagsbuchhandlung, 1869), 3–7; Wilhelm Preyer, *Der Kampf um das Dasein* (Bonn: Weber, 1869), 32–37; *Die Concurrenz in der Natur* (Breslau: S. Schottlaender, 1882), 27–28; Ludwig Woltmann, *Die Darwinsche Theorie und der Sozialismus* (Düsseldorf: Hermann Michels Verlag,

1899), 332 n. For more examples, see Richard Weikart, "The Origins of Social Darwinism in Germany, 1859–1895," *Journal of the History of Ideas* 54 (1993): 469–488.

30. Richard Weikart, *Socialist Darwinism: Evolution in German Socialist Thought from Marx to Bernstein* (San Francisco: International Scholars Publications, 1999), ch. 4.

31. Ludwig Büchner, *Der Mensch und seine Stellung in der Natur in Vergangenheit, Gegenwart und Zukunft*, 2nd ed. (Leipzig: Theodor Thomas, 1872), 198, 201, 208–209; *Darwinismus und Sozialismus oder Der Kampf um das Dasein und die moderne Gesellschaft* (Leipzig: Ernst Günthers Verlag, 1894); for more on Büchner, see Weikart, *Socialist Darwinism*, ch. 3.

32. Hitler, speech to court on first day of trial (February 26, 1924), in *Hitler: Reden, Schriften, Anordnungen: Der Hitler-Prozess 1924: Wortlaut der Hauptverhandlung vor dem Volksgericht München I*, 4 Parts (Munich: K. G. Saur, 1997–1999), part 1: 1–4. *Verhandlungstag* (1997), 21.

33. Hitler, *Mein Kampf*, trans. Ralph Manheim (Boston: Houghton Mifflin, 1943), 214–215.

34. *Hitlers Zweites Buch: Ein Dokument aus dem Jahr 1928*, ed. Gerhard L. Weinberg (Stuttgart: Deutsche Verlags-Anstalt, 1961), ch. 2.

35. Hitler, speech to Second Arbeitskongress of Labor Front on May 16, 1934, in *The Speeches of Adolf Hitler*, 1: 895–896.

36. *Wofür kämpfen wir?* (Berlin: Heerespersonalamt, 1944), 105; emphasis in original.

37. "Appell an die deutsche Kraft," August 4, 1929, in Hitler, *Reden, Schriften, Anordnungen*, vol. 3: 346.

38. R. J. Overy, *War and Economy in the Third Reich* (Oxford: Clarendon Press, 1994), 1–2.

39. Hitler, *Mein Kampf*, 443; emphasis in original.

40. Ernst Haeckel, "Die Wissenschaft und der Umsturz," *Die Zukunft* 10 (1895): 205–206.

41. Hitler, *Mein Kampf*, 446.

42. Ibid., 81.

43. Hitler, *Mein Kampf*, 444–449.

44. "Hitler vor Kreisleitern auf der Ordensburg Volgelsang am 29. April 1937," in *"Es spricht der Führer": 7 exemplarische Hitler-Reden*, ed. Hildegard von Kotze and Helmut Krausnick (Gütersloh: Sigbert Mohn Verlag, 1966), 140–142.

45. Hitler, "Der Wortlaut der Proklamation des Führers," in *Fränkische Tageszeitung* (September 7, 1938), 9, available at Hoover Institution, NSDAP Hauptarchiv, Microfilm Reel 2, Folder 51.

46. Hitler, speech on February 20, 1938, in Domarus, 2: 1024.

47. Hitler, speech to Nuremberg Parteitag, September 1933, in *Speeches of Adolf Hitler*, 1: 476.

48. Hitler, speech on July 1, 1933, in ibid., 1: 483.

49. Heinz Linge, *Bis zum Untergang: Als Chef des Persönlichen Dienstes bei Hitler*, 2nd ed. (Munich: Herbig, 1980), 260.

50. Wagener, *Hitler—Memoirs of a Confidant*, 16; for similar views of other Nazis, see Wolfgang Ayass, *"Asoziale" im Nationalsozialismus* (Stuttgart: Klett-Cotta, 1995), 221–222.

51. August Weismann, "Ueber die Dauer des Lebens," in *Aufsätze über Vererbung und Verwandte Biologische Fragen* (Jena: Gustav Fischer, 1892), 11, 27–28; emphasis in original.

52. Ludwig Büchner, *Die Macht der Vererbung und ihr Einfluss auf den moralischen und geistigen Fortschritt der Menschheit* (Leipzig: Ernst Günthers Verlag, 1882), 100.

53. Wilhelm Schallmayer, *Vererbung und Auslese im Lebenslauf der Völker. Eine Staatswissenschaftliche Studie auf Grund der neueren Biologie* (Jena: Gustav Fischer, 1903), 241–242; Schallmayer, *Beiträge zu einer Nationalbiologie* (Jena: Hermann Costenoble, 1905), 129–131. For further discussion about this issue, see Richard Weikart, *From Darwin to Hitler: Evolutionary Ethics, Eugenics, and Racism in Germany* (New York: Palgrave Macmillan, 2004), ch. 4.

54. Hitler, speech on September 10, 1937, in Domarus, 2: 929.

55. Hitler's speech on May 1, 1939, in ibid., 3: 1603.

56. Hitler speech of May 1, 1934, in ibid., 1: 379.

57. Ibid., 1: 444.

58. Hitler, *Mein Kampf*, 428–431.

59. Bela Bodo, "The Medical Examination and Biological Selection of University Students in Nazi Germany," *Bulletin of the History of Medicine* 76 (2002): 726, 740.
60. "Ansprache des Führeres vor Generalen und Offiziers am 22.6.1944 im Platterhof," in NSDAP Hauptarchiv, Hoover Institution, Microfilm Reel 2, Folder 51.
61. Hitler, *Mein Kampf*, 432–433; emphasis in original.
62. Richard J. Evans, *The Third Reich in Power* (New York: Penguin, 2005), 4, 259, 515.
63. Hitler, *Mein Kampf*, 29.
64. Edward Ross Dickinson, *The Politics of German Child Welfare from the Empire to the Federal Republic* (Cambridge, MA: Harvard University Press, 1996), 222–223.
65. Ayass, *"Asoziale" im Nationalsozialismus*, 20–23.
66. Ibid., 143–165.
67. Götz Aly and Karl Heinz Roth, *The Nazi Census: Identification and Control in the Third Reich*, trans. Edwin Black and Assenka Oksiloff (Philadelphia: Temple University Press, 2004), 99–113.
68. Ayass, *"Asoziale" im Nationalsozialismus*, 105, 206–207.
69. Wolfgang Ayass, ed., *"Gemeinschaftsfremde": Quellen zur Verfolgung von "Asozialen" 1933–1945* (Koblenz: Bundesarchiv, 1998), 253–254.
70. Hitler, "Zukunft oder Untergang" (March 6, 1927), in NSDAP Hauptarchiv, Hoover Institution, Microfilm Reel 2, Folder 59, 18–19.
71. *Adolf Hitler spricht: Ein Lexikon des Nationalsozialismus* (Leipzig: R. Kittler Verlag, 1934), 28.
72. Weikart, "The Origins of Social Darwinism."

Chapter 6

1. Atina Grossmann, *Reforming Sex: The German Movement for Birth Control and Abortion Reform, 1920–1950* (New York: Oxford University Press, 1995), 146.
2. A good balanced approach to Nazi sexual morality is Dagmar Herzog, *Sex after Fascism: Memory and Morality in Twentieth-Century Germany* (Princeton: Princeton University Press, 2005), ch. 1.
3. Hitler, speech on November 23, 1937, in *Hitler: Speeches and Proclamations, 1932–1945*, ed. Max Domarus, 4 vols. (Wauconda, IL: Bolchazy-Carducci Publishers, 1990), 2: 980.
4. Michael Burleigh and Wolfgang Wippermann, *The Racial State: Germany, 1933–1945* (Cambridge: Cambridge University Press, 1991), 193.
5. Jill Stephenson, *Women in Nazi Society* (New York: Barnes and Noble, 1975), 70. The same point is made by Claudia Koonz, *Mothers in the Fatherland* (New York: St. Martin's Press, 1987), 178–180, 408. George L. Mosse, *Nationalism and Sexuality: Respectability and Abnormal Sexuality in Modern Europe* (New York: Howard Fertig, 1985), however, emphasizes the bourgeois respectability of Nazi sexual morality, while also noting its unconventional side.
6. This is also the view of Herzog, *Sex after Fascism*, 17; and Gabriele Czarnowski, *Das kontrollierte Paar: Ehe- und Sexualpolitik im Nationalsozialismsus* (Weinheim: Deutscher Studien Verlag, 1991), 15.
7. Joseph Goebbels, *Die Tagebücher von Joseph Goebbels*, part 1, vol. 9: *December 1940–July 1941*, ed. Elke Fröhlich (Munich: K. G. Saur, 1987ff.), 45–46.
8. Annette Timm, "The Ambivalent Outsider: Prostitution, Promiscuity, and VD Control in Nazi Berlin," in *Social Outsiders in Nazi Germany*, ed. Robert Gellately and Nathan Stolzfus (Princeton: Princeton University Press, 2001), 192–211.
9. *Hitlers Zweites Buch: Ein Dokument aus dem Jahr 1928*, ed. Gerhard L. Weinberg (Stuttgart: Deutsche Verlags-Anstalt, 1961), 46. Hitler made the same point in "Was ist Nationalsozialismus?" (August 6, 1927), and "Tageskampf oder Schicksalskampf" (March 3,

1928), in *Hitler: Reden, Schriften, Anordnungen, Februar 1925 bis Januar 1933* (Munich: K. G. Saur, 1992–2003), vol. 2, part 2: 439, 723.

10. Hitler, "Weltanschauung und Kommunalpolitik" (November 29, 1929), in *Hitler: Reden, Schriften, Anordnungen*, vol. 3, part 2: 488–489.

11. Hitler, "Zukunft oder Untergang," (March 6, 1927), 2, in NSDAP Hauptarchiv, Hoover Institution, Microfilm Reel 2, Folder 59.

12. Hitler, *Mein Kampf*, trans. Ralph Manheim (Boston: Houghton Mifflin, 1943), 405; emphasis in original.

13. Ibid., 403–404.

14. Friedrich Christian Prinz zu Schaumburg-Lippe, *Als die golden Abendsonne: Aus meinen Tagebüchern der Jahre 1933–1937* (Wiesbaden: Limes Verlag, 1971), 114–115.

15. Hitler monologue, January 28, 1942, in *Hitlers Tischgespräche im Führerhauptquartier*, ed. Henry Picker (Frankfurt: Ullstein, 1989), 99–100.

16. Hitler, *Mein Kampf*, 131.

17. Ibid., 132.

18. Hitler made this same point in "Rede auf NSDAP-Parteitag in Nürnberg," August 21, 1927, in *Hitler: Reden, Schriften, Anordnungen*, vol. 2, part 2: 490–492.

19. Hitler, speech of February 15, 1942, in *"Es spricht der Führer": 7 exemplarische Hitler-Reden*, ed. Hildegard von Kotze and Helmut Krausnick (Gütersloh: Sigbert Mohn Verlag, 1966), 312–313. Hitler made the same point again in "War der Zweite Weltkrieg für Deutschland vermeidbar?" (secret speech on May 30, 1942), in *Hitlers Tischgespräche*, 496.

20. Hitler, "Ein Kampf um Deutschlands Freiheit" (February 5, 1928), in *Hitler: Reden, Schriften, Anordnungen*, vol. 2, part 2: 666.

21. Hitler, "Appell an die deutsche Kraft" (August 4, 1929), in ibid., vol. 3, part 2: 348.

22. *Hitler's Second Book: The Unpublished Sequel to* Mein Kampf, ed. Gerhard L. Weinberg (New York: Enigma Books, 2003), 14; see also Hitler, *Monologe im Führerhauptquartier 1941–1944: Die Aufzeichnungen Heinrich Heims*, ed. Werner Jochmann (Hamburg: Albrecht Knaus, 1980), 58.

23. Hitler, *Reden an die deutsche Frau* (Berlin: Schadenverhütung Verlagsgesellschaft, n.d.), 4, 6; emphasis in original.

24. Max von Gruber, *Ursachen und Bekämpfung des Geburtenrückgangs im Deutschen Reich*, 3rd ed. (Munich: Lehmann, 1914).

25. Max von Gruber, "Rassenhygiene, die wichtigste Aufgabe völkischer Innenpolitik," *Deutschlands Erneuerung* 2 (1918): 17–32.

26. Alfred Ploetz, "Neo-Malthusianism and Race Hygiene," in *Problems in Eugenics: Report of Proceedings of the First International Eugenics Congress*, vol. 2 (London: Eugenics Education Society, 1913), 183ff.

27. See Heinz Linge, *Bis zum Untergang: Als Chef des Persönlichen Dienstes bei Hitler*, 2nd ed. (Munich: Herbig, 1980), 85, on Hitler's lack of interest in colleague's sexual affairs.

28. Hitler, *Mein Kampf*, 246–252.

29. Ibid., 252.

30. Richard Weikart, *From Darwin to Hitler: Evolutionary Ethics, Eugenics, and Racism in Germany* (New York: Palgrave Macmillan, 2004), ch. 7.

31. Hitler, *Mein Kampf*, 255; Sebastian Haffner, *The Meaning of Hitler*, trans. Ewald Osers (Cambridge, MA: Harvard University Press, 1983), 36, accuses the Nazis of inconsistency in their sexual morality, but here Hitler attacked eroticism and prudery simultaneously.

32. Lisa Pine, *Nazi Family Policy, 1933–1945* (Oxford: Berg, 1997), 17–18.

33. Ibid., 109; another good discussion of Nazi marriage policy is Michelle Mouton, *From Nurturing the Nation to Purifying the Volk: Weimar and Nazi Family Policy, 1918–1945* (Cambridge: Cambridge University Press, 2007), ch. 1.

34. Hitler, speech of January 1, 1937, in Domarus, 2: 874.

35. Gruber, *Ursachen und Bekämpfung des Geburtenrückgangs*, 63–64.

36. Richard J. Evans, *The Third Reich in Power* (New York: Penguin, 2005), 517.
37. Hitler, *Mein Kampf*, 402–403.
38. Grossmann, *Reforming Sex*, 136–137, 145–146, 151–152; Stephenson, *Women in Nazi Society*, ch. 3.
39. Mosse, *Nationalism and Sexuality*, 158.
40. Grossmann, *Reforming Sex*, 151.
41. Burleigh and Wippermann, *Racial State*, 197.
42. "Erlass des Führers zur Reinhaltung von SS und Polizei," November 15, 1941, in *"Führer-Erlasse" 1939–1945*, ed. Martin Moll (Stuttgart: Franz Steiner Verlag, 1997), 206–207.
43. Joseph Goebbels, August 19, 1941, in *Tagebücher*, part 2, vol. 1: 272.
44. Quoted in Domarus, 1: 567, n. 91.
45. Hitler, April 7, 1942, in *Hitlers Tischgespräche*, 201–203.
46. Edward Ross Dickinson, *The Politics of German Child Welfare from the Empire to the Federal Republic* (Cambridge, MA: Harvard University Press, 1996), 236.
47. Stephenson, *Women in Nazi Society*, 43.
48. Hitler, April 24, 1942 and May 12, 1942, in *Hitlers Tischgespräche*, 241, 288–289.
49. For more on Nazi divorce policy, see Mouton, *From Nurturing the Nation to Purifying the Volk*, ch. 2.
50. Hitler, April 23, 1942, in *Hitlers Tischgespräche*, 235.
51. Mosse, *Nationalism and Sexuality*, 160.
52. Dagmar Herzog, *Sex after Fascism: Memory and Morality in Twentieth-Century Germany* (Princeton: Princeton University Press, 2005), 46.
53. Georg Lilienthal, *Der "Lebensborn e.V."*: *Ein Instrument nationalsozialistischer Rassenpolitik* (Frankfurt: Fischer Taschenbuch Verlag, 1993), 42–43, 90.
54. Richard Breitman, *The Architect of Genocide: Himmler and the Final Solution* (New York: Alfred A. Knopf, 1991), 108.
55. Quoted in Herzog, *Sex after Fascism*, 51.
56. Breitman, *Architect of Genocide*, 109.
57. Lilienthal, *Lebensborn*, 132.
58. Stephenson, *Women in Nazi Society*, 67.
59. Lilienthal, *Lebensborn*, 133.
60. Schaumburg-Lippe, *Als die golden Abendsonne*, 114.
61. Hitler, May 12, 1942, in *Hitlers Tischgespräche*, 288–289.
62. Hitler, March 1, 1942, in *Hitlers Tischgespräche*, 117–118.
63. Felix Kersten, *The Kersten Memoirs, 1940–1945*, trans. Constantine Fitzgibbon and James Oliver (New York: Macmillan, 1957), 176–178, 180.
64. Stephenson, *Women in Nazi Society*, 50–51.

Chapter 7

1. Walter Gross, Vorwort to *Volk und Rasse*, 3rd ed. (Munich: Zentralverlag der NSDAP, Franz Eher Nachf., n.d.), 3.
2. *Wofür kämpfen wir?* (Berlin: Heerespersonalamt, 1944), 4–5, 70, 72, 85, 87, 105, 110.
3. Jill Stephenson, *Women in Nazi Society* (New York: Barnes and Noble, 1975), 62.
4. Hitler, *Mein Kampf*, trans. Ralph Manheim (Boston: Houghton Mifflin, 1943), 249; emphasis in original. I have modified the translation slightly to make it more literal.
5. Ibid., 285. I substituted "evolution" as the proper translation for "Entwicklung" here because context demands it.
6. Ibid., 286–287.

7. Ibid., 286.

8. Ibid., 286, 291–292, 295–296; quote at 289.

9. Ibid., quotes at 283, 325–326.

10. Ibid., 401–402.

11. Ibid., 402; emphasis in original. I have properly translated "Entwicklung" as "evolution" here.

12. Eugen Fischer, *Die Rehobother Bastards und das Bastardierungsproblem beim Menschen* (Jena: Gustav Fischer, 1913), 296–306, quote at 302. For more on Fischer, see Niels Lösch, *Rasse als Konstrukt: Leben und Werk Eugen Fischers* (Frankfurt: Lang, 1997).

13. Hans F. K. Günther, *Rassenkunde des deutschen Volkes*, 3rd ed. (Munich: J. F. Lehmanns Verlag, 1923), 242–243, 502.

14. Hitler, "Interview with George Sylvester Viereck," published in *The American Monthly*, October 1923, in *Hitler: Sämtliche Aufzeichnungen, 1905–1924*, ed. Eberhard Jäckel (Stuttgart: Deutsche Verlags-Anstalt, 1980), 1025.

15. Wilhelm Stuckart and Hans Globke, *Reichsbürgergesetz vom 15. September 1935, Gesetz zum Schutze des deutschen Blutes und der deutschen Ehre vom 15. September 1935, Gesetz zum Schutze der Erbgesundheit des deutschen Volkes (Ehegesundheitsgesetz) vom 18. Oktober 1935 nebst allen Ausführungsvorschriften und den einschlägigen Gesetzen und Verordnungen* (Munich: C. H. Beck'sche Verlagsbuchhandlung, 1936), 263.

16. Jeremy Noakes, "The Development of Nazi Policy towards the German-Jewish 'Mischlinge' 1933–1945," *Leo Baeck Institute Yearbook* 34 (1989): 298.

17. See Eric Ehrenreich, *The Nazi Ancestral Proof: Genealogy, Racial Science, and the Final Solution* (Bloomington: Indiana University Press, 2007), on the practical implementation of Nazi racial laws.

18. Cornelia Essner, *Die "Nürnberger Gesetze" oder Die Verwaltung des Rassenwahns 1933–1945* (Paderborn: Ferdinand Schöningh, 2002), 156–163.

19. Noakes, "Development of Nazi Policy towards the German-Jewish 'Mischlinge,'" 314.

20. Essner, *Nürnberger Gesetze*, 156–163. The Nuremberg Laws and the subsequent decrees regulating them (November 14, 1935) are reprinted in Alexandra Przyrembel, *'Rassenschande': Reinheitsmythos und Vernichtungslegitimation im Nationalsozialismus* (Göttingen: Vandenhoek und Ruprecht, 2003); see also James F. Tent, *In the Shadow of the Holocaust: Nazi Persecution of Jewish-Christian Germans* (Lawrence: University of Kansas Press, 2003), 5–11.

21. Stuckart and Globke, *Reichsbürgergesetz*, 17.

22. *Hitler's Second Book: The Unpublished Sequel to* Mein Kampf, ed. Gerhard L. Weinberg (New York: Enigma Books, 2003), 110–111.

23. Essner, *Nürnberger Gesetze*, 158.

24. Hitler's remarks, December 1, 1941, in *Hitlers Tischgespräche im Führerhauptquartier*, ed. Henry Picker (Frankfurt: Ullstein, 1989), 79.

25. Hitler's remarks, May 10, 1942 and July 1, 1942, in *Hitlers Tischgespräche*, 277, 398–400.

26. Tent, *In the Shadow of the Holocaust*, 2.

27. See Hitler's remarks on July 5, 1942, in *Hitlers Tischgespräche*, 422.

28. Przyrembel, *Rassenschande*, 129.

29. Essner, *Nürnberger Gesetze*, 91–92.

30. Stuckart and Globke, *Reichsbürgergesetz*, 136–137, 153; quote at 38.

31. Hitler, *Mein Kampf*, 624.

32. Reiner Pommerin, *"Sterilisierung der Rheinlandbastarde": Das Schicksal einer farbigen deutschen Minderheit 1918–1937* (Düsseldorf: Droste Verlag, 1979), 49–50, 71–78.

33. David Clay Large, "'Darktown Parade': African Americans in the Berlin Olympics of 1936," *Historically Speaking* 9, 2 (November/December 2007): 7.

34. Vojtech Mastny, *The Czechs under Nazi Rule: The Failure of National Resistance, 1939–1942* (New York: Columbia University Press, 1971), 135.

35. Chad Bryant, *Prague in Black: Nazi Rule and Czech Nationalism* (Cambridge, MA: Harvard University Press, 2007), 163.

36. Ulrich Herbert, *Fremdarbeiter: Politik und Praxis des "Ausländer-Einsatzes" in der Kriegswirtschaft des Dritten Reiches* (Berlin: J. H. W. Dietz Nachf., 1985), 79.

37. Hitler, "Talk with Bormann, Frank, and Schirach in Berlin," October 2, 1940, in *Hitler: Speeches and Proclamations, 1932–1945,* ed. Max Domarus, 4 vols. (Wauconda, IL: Bolchazy-Carducci Publishers, 1990), 3: 2100.

38. Joseph Goebbels, February 20, 1942, in *Die Tagebücher von Joseph Goebbels,* ed. Elke Fröhlich (Munich: K. G. Saur, 1987ff.), part 2, vol. 3: 345.

39. Diemut Majer, *"Non-Germans" under the Third Reich: The Nazi Judicial and Administrative System in Germany and Occupied Eastern Europe with Special Regard to Occupied Poland, 1939–1945,* trans. Peter Thomas Hill et al. (Baltimore: Johns Hopkins University Press, 2003), 209.

40. Robert Gellately, *Backing Hitler: Consent and Coercion in Nazi Germany* (Oxford: Oxford University Press, 2001), 113.

41. Herbert, *Fremdarbeiter,* 75–80.

42. "Erlass des Führers über die Betreuung der unehelichen Kinder von Deutschen in den besetzten Ostgebieten," October 11, 1943, in *"Führer-Erlasse" 1939–1945,* ed. Martin Moll (Stuttgart: Franz Steiner Verlag, 1997), 363.

43. Herbert, *Fremdarbeiter,* 249.

44. Majer, *"Non-Germans" under the Third Reich,* 105.

45. "Erlass des Führers und Reichskanzlers über die Ehen der Beamten des Auswärtigen Dienstes," September 21, 1940, in *Führer-Erlasse,* 138–139.

46. Hitler's Remarks, April 24, 2942, in *Hitlers Tischgespräche,* 240.

47. Joseph Goebbels, March 11, 1941, in *Tagebücher,* part 1, vol. 9: 181; Hitler, September 6, 1942, in *Monologe im Führerhauptquartier 1941–1944: Die Aufzeichnungen Heinrich Heims,* ed. Werner Jochmann (Hamburg: Albrecht Knaus, 1980), 392.

48. Hans-Christian Harten, *De-Kulturation und Germanisierung: Die nationalsozialistische Rassen- und Erziehungspolitik in Polen 1939–1945* (Frankfurt: Campus Verlag, 1996), 90.

49. Hitler's Remarks, July 22, 1942, in *Hitlers Tischgespräche,* 453.

50. Alfred Ploetz to Carl Hauptmann, January 14, 1892, in Carl Hauptmann papers, K 121, Akademie der Künste Archives, Berlin; Alfred Ploetz to Ernst Haeckel, April 18, 1902, in Ernst Haeckel papers, Ernst-Haeckel-Haus, Jena; for more on Ploetz, see W. Doeleke, "Alfred Ploetz (1860–1940): Sozialdarwinist und Gesellschaftsbiologe," dissertation, University of Frankfurt, 1975.

51. Heinrich Ernst Ziegler, "Einleitung zu dem Sammelwerke Natur und Staat," in *Natur und Staat,* vol. 1 (bound with Heinrich Matzat, *Philosophie der Anpassung*) (Jena: Gustav Fischer, 1903), 1–2; Klaus-Dieter Thomann and Werner Friedrich Kümmel, "Naturwissenschaft, Kapital und Weltanschauung: Das Kruppsche Preisausschreiben und der Sozialdarwinismus," *Medizinhistorisches Journal* 30 (1995): 99–143, 205–243. Sheila Faith Weiss provides a good discussion of the Krupp Prize competition in *Race Hygiene and National Efficiency: The Eugenics of Wilhelm Schallmayer* (Berkeley: University of California Press, 1987), 64–74.

52. Wilhelm Schallmayer to Alfred Grotjahn, June 3, 1910, in Alfred Grotjahn papers, Humboldt University Archives, Berlin; Schallmayer makes a similar claim in "Rassedienst," *Sexual-Probleme* 7 (1911): 547.

53. Hitler, "Warum sind wir Antisemiten?" August 13, 1920, in *Hitler: Sämtliche Aufzeichnungen,* 185, 188.

54. Hitler, interview with George Sylvester Viereck, published in *The American Monthly,* October 1923, in ibid., 1025.

55. Hitler, *Mein Kampf,* 29–30.

56. Ibid., 403–405.

57. Ibid., 414.

58. Fritz Lenz, "Die Stellung des Nationalsozialismus zur Rassenhygiene," *Archiv für Rassen- und Gesellschaftsbiologie* 25, 3 (1931): 301–302.

59. Phillip Gassert and Daniel S. Mattern, eds., *The Hitler Library: A Bibliography* (Westport, CT: Greenwood Press, 2001), 48.

60. Günther, *Rassenkunde des deutschen Volkes*, 24, 409–417.

61. Gassert and Mattern, eds., *Hitler Library*, 125.

62. Hitler, "Appell an die deutsche Kraft" (August 4, 1929), in *Hitler: Reden, Schriften, Anordnungen*, vol. 3, part 2: 346–348.

63. "Der Sachverständigen-Beirat für Bevölkerungs- und Rassenpolitik," *Archiv für Rassen- und Gesellschaftsbiologie* 27 (1934): 419–420.

64. Ernst Klee, *Deutsche Medizin im Dritten Reich: Karrieren vor und nach 1945* (Frankfurt: S. Fischer, 2001), 67–68, 127.

65. Robert Proctor, *Racial Hygiene: Medicine under the Nazis* (Cambridge, MA: Harvard University Press, 1988), 108–109; Gisela Bock, "Nazi Sterilization and Reproductive Policies," in *Deadly Medicine: Creating the Master Race* (Washington, DC: United States Holocaust Memorial Museum, 2004), 61–87.

66. Arthur Gütt, Ernst Rüdin, and Falk Ruttke, *Gesetz zur Verhütung erbkranken Nachwuchses vom 14. Juli 1933* (Munich: J. F. Lehmanns Verlag, 1934), 13, quote at 50.

67. Ibid., 174–175; Richard F. Wetzell, *Inventing the Criminal: A History of German Criminology, 1880–1945* (Chapel Hill: University of North Carolina Press, 2000), 257–262.

68. "Das Hamburger Urteil zu Sterilisation und Schwangerschaftsunterbrechung aus eugenischer Indikation," in *Eugenik Sterilisation "Euthanasie": Politische Biologie in Deutschland 1895–1945: Eine Dokumentation*, ed. Jochen-Christoph Kaiser et al. (Berlin: Buchverlag Union, 1992), 134–141.

69. Hans-Walter Schmuhl, *Rassenhygiene, Nationalsozialismus, Euthanasie. Von der Verhütung zur Vernichtung 'lebensunwerten Lebens' 1890–1945* (Göttingen: Vandenhoek und Ruprecht, 1987); 162; Paul Weindling, *Health, Race and German Politics between National Unification and Nazism, 1870–1945* (Cambridge: Cambridge University Press, 1989), 530–531.

70. Proctor, *Racial Hygiene*, 123; Atina Grossmann, *Reforming Sex: The German Movement for Birth Control and Abortion Reform, 1920–1950* (New York: Oxford University Press, 1995), 150.

71. "Vierte Verordnung zur Ausführung des Gesetzes zur Verhütung erbkranken Nachwuchses vom 18. Juli 1935," in *Eugenik Sterilisation "Euthanasie,"* 143.

72. Horst Biesold, *Crying Hands: Eugenics and Deaf People in Nazi Germany*, trans. William Sayers (Washington, DC: Gallaudet University Press, 1999), 84.

73. Hitler, January 30, 1934, in Domarus, 1: 421–422.

74. Biesold, *Crying Hands*, 159.

75. Ulf Schmidt, *Medical Films, Ethics and Euthanasia in Nazi Germany: The History of the Medical Research and Teaching Films of the Reich Office for Educational Films/Reich Institute for Films in Science and Education, 1933–1945* (Husum: Matthiesen Verlag, 2002), 137.

76. Michael Burleigh, *Death and Deliverance: Euthanasia in Germany, 1900–1945* (Cambridge: Cambridge University Press, 1994), ch. 6.

77. Nazi film: *Opfer der Vergangenheit* (1937).

78. Peter Weingart, Jürgen Kroll, and Kurt Bayertz. *Rasse, Blut, und Gene. Geschichte der Eugenik und Rassenhygiene in Deutschland* (Frankfurt: Suhrkamp, 1988), 502–503.

79. Stuckart and Globke, *Reichsbürgergesetz*, 1–30.

80. "Gesetz zum Schutze der Erbgesundheit des deutschen Volkes (Ehegesundheitsgesetz) vom 18. Oktober 1935," *Archiv für Rassen- und Gesellschaftsbiologie* 29 (1935): 361–362; Weingart, Kroll, and Bayertz, *Rasse, Blut, und Gene*, 515–518.

81. Stuckart and Globke, *Reichsbürgergesetz*, 38.

82. "Notizen," *Archiv für Rassen- und Gesellschaftsbiologie* 30 (1936): 285.

83. "Ehrung Alfred Ploetz' durch den Führer," *Archiv für Rassen- und Gesellschaftsbiologie* 29 (1936): frontispiece.
84. Klee, *Deutsche Medizin im Dritten Reich*, 330.
85. "Notizen," *Archiv für Rassen- und Gesellschaftsbiologie*, 29 (1935): 245.
86. Weingart, Kroll, and Bayertz, *Rasse, Blut, und Gene*, 438–439.
87. Stephenson, *Women in Nazi Society*, 69.
88. Hitler, speech on September 7, 1937, in Domarus, 2: 925. I altered the first sentence of this translation slightly by rendering "Rassenhygiene" as "race hygiene."

Chapter 8

1. Hitler, private speech of May 23, 1939, in *Hitler: Speeches and Proclamations, 1932–1945*, ed. Max Domarus, 4 vols. (Wauconda, IL: Bolchazy-Carducci Publishers, 1990), 3: 1618–1619.
2. Hitler, speech of November 23, 1939, in Domarus, 3: 1885.
3. Alex J. Kay, *Exploitation, Resettlement, Mass Murder: Political and Economic Planning for German Occupation Policy in the Soviet Union, 1940–1941* (New York: Berghahn Books, 2006), 4–7, 121, takes this mistaken position.
4. Gerhard Weinberg, *The Foreign Policy of Hitler's Germany*, vol. 1: *Diplomatic Revolution in Europe, 1933–36* (Chicago: University of Chicago Press, 1970), 6. Wendy Lower makes a similar argument in *Nazi Empire-Building and the Holocaust in Ukraine* (Chapel Hill: University of North Carolina Press, 2005).
5. Wolfram Wette, "Ideology, Propaganda, and Internal Politics as Preconditions of the War Policy of the Third Reich," in *Germany and the Second World War*, ed. Militärgeschichtliches Forschungsamt, Freiburg, vol. 1: *The Build-up of German Aggression* (Oxford: Clarendon Press, 1990), 18–20.
6. Karl Haushofer, "Friedrich Ratzel als raum- und volkspolitischer Gestalter," in Friedrich Ratzel, *Erdenmacht und Völkerschicksal. Eine Auswahl aus seinen Werken*, ed. Karl Haushofer (Stuttgart: Alfred Kröner, 1940), xxv–xxvi.
7. Hans-Adolf Jacobsen, *Karl Haushofer: Leben und Werk*, vol. 1: *Lebensweg 1869–1946 und ausgewählte Texte zur Geopolitik* (Boppard am Rhein: Harald Boldt Verlag, 1979), 239.
8. Othmar Plöckinger, *Geschichte eines Buches: Adolf Hitlers "Mein Kampf" 1922–1945* (Munich: R. Oldernbourg, 2006), 52–53.
9. Bruno Hipler, *Hitlers Lehrmeister: Karl Haushofer als Vater der NS-Ideologie* (St. Ottilien: EOS-Verlag, 1996), 44.
10. See Eberhard Jäckel, *Hitler's Weltanschauung: A Blueprint for Power*, trans. Herbert Arnold (Middleton, CT: Wesleyan University Press, 1972), ch. 2, on the changes in Hitler's foreign policy in his early years. Michael Kellogg, *The Russian Roots of Nazism: White Émigrés and the Making of National Socialism, 1917–1945* (Cambridge: Cambridge University Press, 2005), 167, also discusses this issue.
11. Hitler, "Gespräch mit Eduard August Scharrer" (end of December 1922), in *Hitler: Sämtliche Aufzeichnungen, 1905–1924*, ed. Eberhard Jäckel (Stuttgart: Deutsche Verlags-Anstalt, 1980), 773.
12. Hitler, speech of December 10, 1919, in Reginald H. Phelps, "Hitler als Parteiredner," *Vierteljahrshefte für Zeitgeschichte*," 11 (1963): 289.
13. Report on Hitler, "Der Arbeiter im Deutschland der Zukunft" (November 19, 1920), in *Hitler: Sämtliche Aufzeichnungen*, 259.
14. Geoffrey Stoakes, *Hitler and the Quest for World Dominion* (Leamington Spa, UK: Berg, 1986), 54–55, considers the former interpretation more likely.

15. Hitler: *Reden, Schriften, Anordnungen. Der Hitler-Prozess 1924: Wortlaut der Hauptverhandlung vor dem Volksgericht München I.* Part 4: *19.-25.Verhandlungstag* (Munich: K. G. Saur, 1997–1999), 1582–1583.

16. Hitler, *Mein Kampf*, trans. Ralph Manheim (Boston: Houghton Mifflin, 1943), 3.

17. Ibid., 131–140.

18. Ibid., 134–140; quote at 134.

19. Ibid., 642–643, 646, 649; quote at 652–653.

20. Ibid., 654–662.

21. Hitler, "Die soziale Sendung des Nationalsozialismus" (Flugblatt, December 16, 1925), in *Hitler: Reden, Schriften, Anordnungen, Februar 1925 bis Januar 1933* (Munich: K. G. Saur, 1992–1995), 1: 240–242, 258; emphasis in original. Hitler made the same point about emigration in *Hitler's Second Book: The Unpublished Sequel to* Mein Kampf, ed. Gerhard L. Weinberg (New York: Enigma Books, 2003), 13–14.

22. For examples, see Hitler, "20 Millionen Deutsche zuviel?" (May 7, 1927), "Was ist Nationalsozialismus?" (August 6, 1927), and "Tageskampf oder Schicksalskampf," (March 3, 1928), in *Hitler: Reden, Schriften, Anordnungen*, 2: 292, 442–445, 723.

23. Hitler, "Appell an die deutsche Kraft" (August 4, 1929), in *Hitler: Reden, Schriften, Anordnungen*, 3: 347–348.

24. *Hitler's Second Book*, 8–18; quotes at 8, 18.

25. Richard Bessel, *Nazism and War* (New York: Modern Library, 2004), xv.

26. *Hitler's Second Book*, 18–20, 158.

27. Reinhard Müller, "Hitlers Rede vor der Reichswehrführung 1933: Eine neue Moskauer Überlieferung," *Mittelweg* 36, 1 (2001): 77–78.

28. Hitler, interview with Ward Price (October 18, 1933), in Domarus, 1: 382.

29. Hitler, speech on May 1, 1937, in Domarus, 2: 891–892.

30. Hitler, speeches on October 3, 1937 and November 21, 1937, in Domarus, 2: 952, 978.

31. Hitler, speech on May 1, 1939, in Domarus, 3: 1604. In the translation of Domarus, "für uns das höchste Gebot" is translated as "the command of the hour." I have translated it more literally.

32. "Adolf Hitlers Geheimrede vom 23. November 1937 auf der Ordensburg Sonthofen," in *Hitlers Tischgespräche im Führerhauptquartier*, ed. Henry Picker (Frankfurt: Ullstein, 1989), 481–490; quote at 486; emphasis in original.

33. Hitler, "Denkschrift des Führers," (1936), in Records of Beauftragten für den Vierjahresplan (Göring), RG-11.001M.09, Reel 205, in U.S. Holocaust Memorial Museum Archives.

34. R. J. Overy, *War and Economy in the Third Reich* (Oxford: Clarendon Press, 1994), 194–195, 188–189, 226–227; quote at 235.

35. Hitler, briefing to military leaders on May 23, 1939, in Domarus, Eng., 3: 1620.

36. Overy, *War and Economy in the Third Reich*, 309–310.

37. Hossbach Memorandum, at www.yale.edu/lawweb/avalon/imt/hossbach.htm, accessed October 3, 2001.

38. Hitler, *Mein Kampf*, 405.

39. *Hitler's Second Book*, 53.

40. Alexander Rossino, *Hitler Strikes Poland: Blitzkrieg, Ideology, and Atrocity* (Lawrence: University Press of Kansas, 2003), 3.

41. Uwe Mai, *"Rasse und Raum": Agrarpolitik, Sozial-und Raumplanung im NS-Staat* (Paderborn: Ferdinand Schöningh, 2002), 78–79.

42. Chad Bryant, *Prague in Black: Nazi Rule and Czech Nationalism* (Cambridge, MA: Harvard University Press, 2007), 3, 104–117.

43. Franz Halder, *Kriegstagebuch*, vol. 1: *Vom Polenfeldzug bis zum Ende der Westoffensive (14.8.1939–30.6.1940)*, ed. Hans-Adolf Jacobsen (Stuttgart: W. Kohlhammer Verlag, 1963), 25–26.

44. Rossino, *Hitler Strikes Poland*, 10–17, 57, 234–235.
45. Joseph Goebbels, September 30, 1939, *Die Tagebücher von Joseph Goebbels*, part 1, vol. 7 *(Juli 1939–März 1940)*, ed. Elke Fröhlich (Munich: K. G. Saur, 1987ff.), 130.
46. Phillip T. Rutherford, *Prelude to the Final Solution: The Nazi Program for Deporting Ethnic Poles, 1939–1941* (Lawrence: University Press of Kansas, 2007), 9.
47. Saul Friedländer, *The Years of Extermination: Nazi Germany and the Jews, 1939–1945* (New York: Harper Perennial, 2007), 11.
48. Mai, *Rasse und Raum*, 153.
49. Goebbels, March 13, 1940, *Tagebücher*, part 1, vol. 7: 346–347.
50. Martyn Housden, *Hans Frank: Lebensraum and the Holocaust* (Houndmills: Palgrave Macmillan, 2003), 96.
51. Hitler, briefing to military leaders on May 23, 1939, in Domarus, 3: 1619.
52. Goebbels, November 5, 1940, *Tagebücher*, part 1, vol. 7: 406.
53. Jan Tomasz Gross, *Polish Society under German Occupation: The Generalgouvernement, 1939–1944* (Princeton: Princeton University Press, 1979), 72–73.
54. Ibid., 49, 75–76; Gross claims that Nazi long-term policy was extermination for Poles.
55. Raffael Scheck, *Hitler's African Victims: The German Army Massacres of Black French Soldiers in 1940* (Cambridge: Cambridge University Press, 2006), 3–9, 60, 102–106, 117, 165.
56. "Die Garde der Zivilisation," *Das schwarze Korps* (June 6, 1940), 8. Thanks to Raffael Scheck for helping me locate this quotation.
57. Hitler monologue, September 23, 1941, in *Monologe im Führerhauptquartier 1941–1944: Die Aufzeichnungen Heinrich Heims*, ed. Werner Jochmann (Hamburg: Albrecht Knaus, 1980), 66.
58. Hitler monologue, October 10, 1941, in ibid., 76.
59. Franz Halder, March 30, 1941, in *Kriegstagebuch*, 2: 336–337.
60. Geoffrey Megargee, *War of Annihilation: Combat and Genocide on the Eastern Front, 1941* (Lanham, MD: Rowman and Littlefield, 2006), 33.
61. Ibid., 34.
62. Karel C. Berkhoff, *Harvest of Despair: Life and Death in Ukraine under Nazi Rule* (Cambridge, MA: Belknap Press of Harvard University Press, 2004), 35, 164–165.
63. Goebbels, July 9, 1941, August 19, 1941, September 24, 1941, November 22, 1941, and November 30, 1941, in *Tagebücher*, part 2, vol. 1: 33, 260–261, 481–482; 2: 337, 399.
64. Goebbels, November 30, 1941 and December 13, 1941, *Tagebücher*, part 2, vol. 2: 400–401, 499.
65. Hitler monologue, May 12, 1942, in *Hitlers Tischgespräche*, 284.
66. Goebbels, May 24, 1942, *Tagebücher*, part 2, vol. 4: 362–363.
67. Christopher Browning, with contributions by Jürgen Matthäus, *The Origins of the Final Solution: The Evolution of Nazi Jewish Policy, September 1939–March 1942* (Lincoln: University of Nebraska Press, 2004), 108.
68. Berkhoff, *Harvest of Despair*, 44–45. Wendy Lower, "A New Ordering of Space and Race: Nazi Colonial Dreams in Zhytomyr, Ukraine, 1941–1944," *German Studies Review* 25 (2002): 227–254, also discusses the Nazi colonization plans in Ukraine.
69. The five I do not analyze (November 23, 1939; May 3, 1940; December 18, 1940; April 29, 1941; and May 30, 1942) are in Domarus, 3: 1884–1890, 1978–1988, 2160–2171, 4: 2410; and "Adolf Hitlers Geheimrede vor dem 'Militärischen Führernachwuchs' vom 30. Mai 1942, 'War der Zweite Weltkrieg für Deutschland vermeidbar?'" in *Hitlers Tischgespräche*, 491–502.
70. Hitler, "Ansprache des Führeres vor Generalen und Offiziers am 22.6.1944 im Platterhof," in Hoover Institution, NSDAP Hauptarchiv, Reel 2, Folder 51.
71. Hitler, speech of February 15, 1942, in *"Es spricht der Führer": 7 exemplarische Hitler-Reden*, ed. Hildegard von Kotze and Helmut Krausnick (Gütersloh: Sigbert Mohn Verlag, 1966), 312–313.

72. Ibid., 308.
73. Megargee, *War of Annihilation*, 4, see also 8–9.
74. Omer Bartov, *Hitler's Army: Soldiers, Nazis, and War in the Third Reich* (Oxford: Oxford University Press, 1991).
75. Hitler, *Mein Kampf*, 520–530; Hitler repeated this theme in "Rede vor dem Nationalklub von 1919 in Hamburg," February 28, 1926, in *Hitler: Reden, Schriften, Anordnungen*, vol. 1: 303–304.
76. *Hitler's Second Book*, 11–12.
77. Robert Proctor, *Racial Hygiene: Medicine under the Nazis* (Cambridge, MA: Harvard University Press, 1988), 28–29.
78. Hitler, monologues of May 22, 1942 and May 29, 1942, in *Hitlers Tischgespräche*, 331–332, 338.
79. Goebbels, September 23, 1941, *Tagebücher*, part 2, vol. 1: 474.
80. Hitler, monologue on January 28, 1942, in *Hitlers Tischgespräche*, 99–100.
81. "Adolf Hitlers Geheimrede vor dem 'Militärischen Führernachwuchs' vom 30. Mai 1942, 'War der Zweite Weltkrieg für Deutschland vermeidbar?,'" in *Hitlers Tischgespräche*, 496.
82. Hitler monologue on August 19–20, 1941, in *Monologe im Führerhauptquartier*, 58.

Chapter 9

1. Bettina Winter, ed., *"Verlegt nach Hadamar": Die Geschichte einer NS-"Euthanasie"-Anstalt* (Kassel: Landeswolfahrtsverband Hessen, 1991), 69.
2. Joseph Goebbels, May 1, 1940, *Die Tagebücher von Joseph Goebbels*, ed. Elke Fröhlich (Munich: K. G. Saur, 1987ff.), part 1, vol. 8: 87.
3. Henry Friedlander, *The Origins of Nazi Genocide: From Euthanasia to the Final Solution* (Chapel Hill: University of North Carolina Press, 1995), ch. 1; Michael Burleigh, *Death and Deliverance: Euthanasia in Germany, 1900–1945* (Cambridge: Cambridge University Press, 1994); Hans-Walter Schmuhl, *Rassenhygiene, Nationalsozialismus, Euthanasie. Von der Verhütung zur Vernichtung 'lebensunwerten Lebens' 1890–1945* (Göttingen: Vandenhoek und Ruprecht, 1987); Ernst Klee, *"Euthanasie" im NS-Staat: Die "Vernichtung lebensunwerten Lebens"* (Frankfurt: Fischer Taschenbuch, 1985); Peter Weingart, Jürgen Kroll, and Kurt Bayertz, *Rasse, Blut, und Gene: Geschichte der Eugenik und Rassenhygiene in Deutschland* (Frankfurt: Suhrkamp, 1988), 523–524.
4. Schmuhl, *Rassenhygiene, Nationalsozialismus, Euthanasie*, quotes at 18–19, 106.
5. Ian Dowbiggin, *A Merciful End: The Euthanasia Movement in Modern America* (Oxford: Oxford University Press, 2003), 8; N. D. A. Kemp, *'Merciful Release': The History of the British Euthanasia Movement* (Manchester: Manchester University Press, 2002), 19.
6. Ernst Haeckel, *Natürliche Schöpfungsgeschichte*, 2nd ed. (Berlin: Georg Reimer, 1870), 152–155.
7. Ernst Haeckel, *Die Lebenswunder: Gemeinverständliche Studien über Biologische Philosophie* (Stuttgart: Alfred Kröner, 1904), 21–22, 33–34, 134–136.
8. August Forel, *Die sexuelle Frage* (Munich: Ernst Reinhardt, 1905), 399–400; see also Forel, *Leben und Tod* (Munich: Ernst Reinhardt, 1908), 3; and *Kulturbestrebungen der Gegenwart* (Munich: Ernst Reinhardt, 1910), 26–27.
9. Carl Vogt, *Vorlesungen über den Menschen, seine Stellung in der Schöpfung und in der Geschichte der Erde*, 2 vols. (Giessen: J. Ricker'sche Buchhandlung, 1863), 1: 214, 256; Andrew Zimmerman, *Anthropology and Antihumanism in Imperial Germany* (Chicago: University of Chicago Press, 2001), 75.
10. See Richard Weikart, *From Darwin to Hitler: Evolutionary Ethics, Eugenics, and Racism in Germany* (New York: Palgrave Macmillan, 2004), ch. 8.

11. Götz Aly and Karl Heinz Roth, *The Nazi Census: Identification and Control in the Third Reich*, trans. Edwin Black and Assenka Oksiloff (Philadelphia: Temple University Press, 2004), 97.

12. Friedlander, *Origins of Nazi Genocide*, xi–xii, 64, 81, argues that economic considerations were subsidiary to ideology; Robert Proctor, *Racial Hygiene: Medicine under the Nazis* (Cambridge, MA: Harvard University Press, 1988), 183–184, on the other hand, argues that economics was a primary motivating factor.

13. Hitler, August 20, 1942, in *Monologe im Führerhauptquartier 1941–1944: Die Aufzeichnungen Heinrich Heims*, ed. Werner Jochmann (Hamburg: Albrecht Knaus, 1980), 348–349.

14. For extended discussion of this point among eugenicists in the Nazi period, see Stefan Kühl, "The Relationship between Eugenics and the So-Called 'Euthanasia Action' in Nazi Germany: A Eugenically Motivated Peace Policy and the Killing of the Mentally Handicapped during the Second World War," in *Science and the Third Reich*, ed. Margit Szöllösi-Janze (Oxford: Berg, 2001), 185–210.

15. Hitler, interview with George Sylvester Viereck, published in *The American Monthly* (October 1923), in *Hitler Sämtliche Aufzeichnungen, 1905–1924*, ed. Eberhard Jäckel (Stuttgart: Deutsche Verlags-Anstalt, 1980), 1025.

16. "Ein Kampf um Deutschlands Freiheit" (February 5, 1928), in *Hitler: Reden, Schriften, Anordnungen, Februar 1925 bis Januar 1933*, 6 vols. (Munich: K. G. Saur, 1992–2003), vol. 2, part 2: 665.

17. *Hitler's Second Book: The Unpublished Sequel to* Mein Kampf, ed. Gerhard L. Weinberg (New York: Enigma Books, 2003), 21.

18. Hitler, "Appell an die deutsche Kraft" (August 4, 1929), in *Hitler: Reden, Schriften, Anordnungen*, 3: 348–349.

19. Otto Wagener, *Hitler—Memoirs of a Confidant*, ed. Henry Ashby Turner, trans. Ruth Hein (New Haven: Yale University Press, 1985), 40.

20. Ibid.

21. Ibid., 145–147.

22. Haeckel, *Lebenswunder*, 22.

23. Wolfgang Eckart and Andreas Reuland, "First Principles: Julius Moses and Medical Experimentation in the Late Weimar Republic," in *Man, Medicine, and the State: The Human Body as an Object of Government Sponsored Medical Research in the 20th Century*, ed. Wolfgang U. Eckart (Stuttgart: Franz Steiner Verlag, 2006), 40.

24. Edward Ross Dickinson, *The Politics of German Child Welfare from the Empire to the Federal Republic* (Cambridge, MA: Harvard University Press, 1996), 233.

25. Proctor, *Racial Hygiene*, 181.

26. Schmuhl, *Rassenhygiene, Nationalsozialismus, Euthanasie*, 180–181.

27. Friedlander, *Origins of Nazi Genocide*, 39–40; Ulf Schmidt, *Medical Films, Ethics and Euthanasia in Nazi Germany: The History of the Medical Research and Teaching Films of the Reich Office for Educational Films/Reich Institute for Films in Science and Education, 1933–1945* (Husum: Matthiesen Verlag, 2002), 242–243.

28. Friedlander, *Origins of Nazi Genocide*, 63.

29. Andreas Frewer, *Medizin und Moral in Weimarer Republik und Nationalsozialismus: Die Zeitschrift "Ethik" unter Emil Abderhalden* (Frankfurt: Campus Verlag, 2000), 97, 119, 127, 154.

30. Friedlander, *Origins of Nazi Genocide*, 71, 78.

31. Hitler, September 30, 1942, in *Hitler: Speeches and Proclamations, 1932–1945*, ed. Max Domarus, 4 vols. (Wauconda, IL: Bolchazy-Carducci Publishers, 1990), 4: 2684.

32. Hitler, August 1, 1942 and August 20, 1942, *Monologe im Führerhauptquartier*, 320, 348–349.

33. Goebbels, May 23, 1942, *Tagebücher*, part 2, vol. 4: 343; Hitler, May 22, 1942, *Hitlers Tischgespräche im Führerhauptquartier*, ed. Henry Picker (Frankfurt: Ullstein, 1989), 331–332; see also Hitler, *Mein Kampf*, trans. Ralph Manheim (Boston: Houghton Mifflin, 1943), 169.

34. Barry W. Butcher, "Darwinism, Social Darwinism and the Australian Aborigines: A Reevaluation," in *Darwin's Laboratory: Evolutionary Theory and Natural History in the Pacific*, ed. Roy MacLeod and Philip F. Rehbock (Honolulu: University of Hawaii Press, 1994), 371–394; Janet Browne, *Charles Darwin*, vol. 1: *Voyaging* (New York: Alfred Knopf, 1995), 421–422.

35. Tony Barta, "Mr. Darwin's Shooters: On Natural Selection and the Naturalizing of Genocide," *Patterns of Prejudice* 39 (2005): 116–137; Barta, "On Pain of Extinction: Laws of History in Darwin, Marx, and Arendt," in *Hannah Arendt and the Uses of History: Imperialism, Nation, Race, and Genocide*, ed. Richard H. King and Dan Stone (New York: Berghahn Books, 2007), 87–105; Adrian Desmond and James Moore, *Darwin* (London: Michael Joseph, 1991), xxi, 521,

36. Francis Darwin, *Charles Darwin: His Life Told in an Autobiographical Chapter, and in a Selected Series of His Published Letters* (London: John Murray, 1902), 64.

37. Richard Weikart, "Progress through Racial Extermination: Social Darwinism, Eugenics, and Pacifism in Germany, 1860–1918," *German Studies Review* 26 (2003): 273–294.

38. Stig Förster and Myriam Gesslerr, "The Ultimate Horror: Reflections on Total War and Genocide," in *A World at Total War: Global Conflict and the Politics of Destruction, 1937–1945*, ed. Roger Chickering, Stig Förster, and Bernd Greiner (Cambridge: Cambridge University Press, 2005), 67.

39. Friedlander, *Origins of Nazi Genocide*, xii–xiii; Proctor, *Racial Hygiene*, 195.

40. Hitler, April 4, 1942, in *Hitlers Tischgespräche*, 186–187.

41. Hitler, "Teuerungsprotest ein Judenschwindel" (August 12, 1921), in *Hitler Sämtliche Aufzeichnungen*, 456.

42. Hitler, "Der Klassenkampf ein Börsenbetrug" (March 1, 1922), in ibid., 588.

43. Hitler, speech on August 7, 1920, in ibid., 176–177. Hitler used this trope quite frequently; see, for example, Hitler, letter to Konstantin Hierl, July 3, 1920, in ibid., 156; Hitler, interview with George Sylvester Viereck, published in *The American Monthly* (October 1923), in ibid., 1025; Hitler, "Warum musste ein 8. November kommen?" *Deutschland Erneuerung* 8 (April 1924): 201.

44. Hitler, "Warum sind wir Antisemiten?" (August 13, 1920), in *Hitler: Sämtliche Aufzeichnungen*, 195.

45. Hitler, "Warum musste ein 8. November kommen?" 201.

46. Hitler, speech on September 14, 1936, in Domarus, 2: 839.

47. Hitler, speech on January 30, 1939, in Domarus, 3: 1449.

48. Christian Gerlach, "The Wannsee Conference, the Fate of the German Jews, and Hitler's Decision in Principle to Exterminate All European Jews," *Journal of Modern History* 70 (1998): 759–812.

49. Goebbels, August 17, 1940, *Tagebücher*, part 1, vol. 8: 276.

50. Hitler, *Mein Kampf*, 302.

51. Goebbels, May 30, 1942, in *Tagebücher*, part 2, vol. 4: 406.

52. Saul Friedländer concurs with this view: Saul Friedländer, *The Years of Extermination: Nazi Germany and the Jews, 1939–1945* (New York: Harper Perennial, 2007), 240.

53. Goebbels, August 19, 1941, in *Tagebücher*, part 2, vol. 1: 269.

54. Friedländer, *Years of Extermination*, 284–285.

55. Goebbels, December 13, 1941, *Tagebücher*, part 2, vol. 2: 498–499.

56. Hitler, "Ansprache des Führers vor Generalen und Offiziers am 22.6.1944 im Platterhof," in Hoover Institution, NSDAP Hauptarchiv, Microfilm Reel 2, Folder 51, pp. 3–4, 38–39.

57. "Ansprache Hitlers vor Generalen und Offizieren am 26. Mai 1944 im Platterhof," in Hans-Heinrich Wilhelm, "Hitlers Ansprache vor Generalen und Offizieren am 26. Mai 1944," *Militärgeschichtliche Mitteilungen* 2 (1976): 146–147, 155–157.

58. Ernst Klee, *Deutsche Medizin im Dritten Reich: Karrieren vor und nach 1945* (Frankfurt: S. Fischer, 2001), 256, 331.

59. Gretchen E. Schafft, *From Racism to Genocide: Anthropology in the Third Reich* (Urbana: University of Illinois Press, 2004).
60. Friedlander, *Origins of Nazi Genocide*, 134–135.
61. Christopher Hutton, *Race and the Third Reich: Linguistics, Racial Anthropology and Genetics in the Dialectic of Volk* (Cambridge, UK: Polity, 2005), 212.

Conclusion

1. Robert Proctor, *Racial Hygiene: Medicine under the Nazis* (Cambridge, MA: Harvard University Press, 1988), 62; H. Linder and R. Lotze, "Lehrplanentwurf für den biologischen Unterricht an den höheren Knabenschulen," in *Der Biologe*, published as separate supplement without page numbering in vol. 6 (1937)—it appeared immediately after Heft 1 in the copy I read at the University of California's Northern Regional Library Facility.
2. *Wofür kämpfen wir?* (Berlin: Heerespersonalamt, 1944), iv–vi.
3. Ibid., 67, 68, 105, 110.
4. Phillip T. Rutherford, *Prelude to the Final Solution: The Nazi Program for Deporting Ethnic Poles, 1939–1941* (Lawrence: University Press of Kansas, 2007); Chad Bryant, *Prague in Black: Nazi Rule and Czech Nationalism* (Cambridge, MA: Harvard University Press, 2007); John Connelly, "Nazis and Slavs: From Racial Theory to Racist Practice," *Central European History* 32 (1999): 1–33.
5. Rutherford, *Prelude to the Final Solution*, 213–216.
6. Hitler, briefing of military leaders on May 23, 1939, in *Hitler: Speeches and Proclamations, 1932–1945*, ed. Max Domarus, 4 vols. (Wauconda, IL: Bolchazy-Carducci Publishers, 1990), 3: 1618–1619.
7. Uwe Mai, *"Rasse und Raum": Agrarpolitik, Sozial-und Raumplanung im NS-Staat* (Paderborn: Ferdinand Schöningh, 2002), 108.
8. Gilmer Blackburn, *Education in the Third Reich: Race and History in Nazi Textbooks* (Albany: State University of New York Press, 1985), 21–22, 65.
9. Gerhard Trommer, "Bezüge des NS-Lebenskunde zur Ökologie," in *Medizin, Naturwissenschaft, Technik und Nationalsozialismus: Kontinuitäten und Diskontinuitäten*, ed. Christoph Meinel and Peter Voswinckel (Stuttgart: Verlag für Geschichte der Naturwissenschaft und der Technik, 1994), 144.
10. The official Nazi biology curriculum is contained in Linder and Lotze, "Lehrplanentwurf für den biologischen Unterricht."
11. *Rassenpolitik* (Berlin: Der Reichsführer SS, SS-Hauptamt, n.d.).
12. Andreas Frewer, *Medizin und Moral in Weimarer Republik und Nationalsozialismus: Die Zeitschrift "Ethik" unter Emil Abderhalden* (Frankfurt: Campus Verlag, 2000).
13. Michael Burleigh, *Death and Deliverance: Euthanasia in Germany, 1900–1945* (Cambridge: Cambridge University Press, 1994), ch. 6.
14. Ulf Schmidt, *Medical Films, Ethics and Euthanasia in Nazi Germany: The History of the Medical Research and Teaching Films of the Reich Office for Educational Films/Reich Institute for Films in Science and Education, 1933–1945* (Husum: Matthiesen Verlag, 2002), 51, 144, 154.
15. Recently Eric Ehrenreich has argued forcefully for this position in "Otmar von Verschuer and the 'Scientific' Legitimization of Nazi Anti-Jewish Policy," *Holocaust and Genocide Studies* 21, 1 (2007): 55–72; see also Ehrenreich, *The Nazi Ancestral Proof: Genealogy, Racial Science, and the Final Solution* (Bloomington: Indiana University Press, 2007).
16. Benno Müller-Hill, "Reflections of a German Scientist," in *Deadly Medicine: Creating the Master Race* (Washington, DC: United States Holocaust Memorial Museum, 2004); see also Müller-Hill, *Murderous Science: Elimination by Scientific Selection of Jews, Gypsies, and Others, Germany, 1933–1945*, trans. George R. Fraser (Oxford: Oxford University Press, 1988).

Benoit Massin makes the same point in "The 'Science of Race,'" in *Deadly Medicine*; and in "Rasse und Vererbung als Beruf: Die Hauptforschungsrichtungen am Kaiser-Wilhlem-Institut für Anthropologie, menschliche Erblehre und Eugenik im Nationalsozialismus," in *Rassenforschung an Kaiser-Wilhelm-Instituten vor und nach 1933*, ed. Hans-Walter Schmuhl (Göttingen: Wallstein Verlag, 2003), 190.

17. Many recent works on Nazi eugenics and anthropology detail this cooperation. One especially good recent work is Hans-Walter Schmuhl, *Grenzüberschreitungen: Das Kaiser-Wilhelm-Institut für Anthropologie, menschliche Erblehre und Eugenik 1927–1945* (Göttingen: Wallstein Verlag, 2005); see also Gretchen E. Schafft, *From Racism to Genocide: Anthropology in the Third Reich* (Urbana: University of Illinois Press, 2004).

18. Richard Overy compares the moral views of Hitler and Stalin in *The Dictators: Hitler's Germany and Stalin's Russia* (New York: W. W. Norton, 2004), 265–267.

BIBLIOGRAPHY

Archival Materials

Akademie der Künste Archives, Berlin,
  Carl Hauptmann papers.
Ernst-Haeckel-Haus Archives, Jena,
  Ernst Haeckel papers.
Hoover Institution,
  NSDAP Hauptarchiv.
Humboldt University Archives, Berlin,
  Alfred Grotjahn papers.
United States Holocuast Memorial Museum,
  Records of Four-Year Plan.
United States Library of Congress,
  Hitler's Personal Library.
University of Freiburg Library Archives,
  Ludwig Schemann papers.

Primary Sources—Journals

Archiv für Rassen- und Gesellschaftsbiologie.
Das Ausland.
Der Biologe.
Deutschlands Erneuerung.
Neues Volk: Blätter des Aufklärungsamt für Bevölkerungspolitik und Rassenpflege.
Ostara.
Politisch-anthropologische Revue.
Das schwarze Korps.

Volk und Rasse.

Die Zukunft.

Primary Sources—Books and Articles

Ayass, Wolfgang, ed. *"Gemeinschaftsfremde": Quellen zur Verfolgung von "Asozialen" 1933–1945.* Koblenz: Bundesarchiv, 1998.

Below, Nicolaus von. *Als Hitlers Adjutant 1937–45.* Mainz: v. Hase and Koehler Verlag, 1980.

Büchner, Ludwig. *Darwinismus und Sozialismus, oder Der Kampf um das Dasein und die moderne Gesellschaft.* Leipzig: Ernst Günthers Verlag, 1894.

———. *Die Macht der Vererbung und ihr Einfluss auf den moralischen und geistigen Fortschritt der Menschheit.* Leipzig: Ernst Günthers Verlag, 1882.

———. *Der Mensch und seine Stellung in der Natur in Vergangenheit, Gegenwart und Zukunft.* 2nd ed. Leipzig: Verlag von Theodor Thomas, 1872.

Chamberlain, Houston Stewart. *Die Grundlagen des neunzehnten Jahrhunderts.* 2 vols. Munich: F. Bruckmann, 1899.

Darwin, Charles. *The Descent of Man.* 2 vols. in 1. London, 1871. Reprint, Princeton: Princeton University Press, 1981.

———. *The Origin of Species.* London: Penguin, 1968.

Darwin, Francis. *Charles Darwin: His Life Told in an Autobiographical Chapter, and in a Selected Series of His Published Letters.* London: John Murray, 1902.

Dietrich, Otto. *The Hitler I Knew.* Trans. Richard and Clara Winston. London: Methuen, 1957.

Domarus, Max, ed. *Hitler. Reden und Proklamationen 1932–1945.* 2 vols. in 4 parts. Munich: Süddeutscher Verlag, 1965.

———, ed. *Hitler: Speeches and Proclamations, 1932–1945.* 4 vols. Wauconda, IL: Bolchazy-Carducci Publishers, 1990.

Eckart, Dietrich. *Der Bolschewismus von Moses bis Lenin. Zwiegespräch zwischen Adolf Hitler und mir.* Munich: Hoheneichen Verlag, 1924.

Fischer, Eugen. *Die Rehobother Bastards und das Bastardierungsproblem beim Menschen.* Jena: Gustav Fischer, 1913.

Forel, August. *Kulturbestrebungen der Gegenwart.* Munich: Ernst Reinhardt, 1910.

———. *Leben und Tod.* Munich: Ernst Reinhardt, 1908.

———. *Die Sexuelle Frage.* Munich: Ernst Reinhardt, 1905.

Frank, Hans. *Im Angesicht des Galgens: Deutung Hitlers und seiner Zeit auf Grund eigener Erlebnisse und Erkenntnisse.* Munich-Gräfelfing: Friedrich Alfred Beck Verlag, 1953.

Fritsch, Theodor. *Mein Streit mit dem Hause Warburg: Eine Episode aus dem Kampfe gegen das Weltkapital.* Leipzig: Hammer-Verlag, 1925.

Fünfzig Jahre J. F. Lehmanns *Verlag 1890–1940: Zur Erinnerung an das fünfzigjährige Bestehen am 1. September 1940.* Munich: J. F. Lehmanns Verlag, 1940.

Goebbels, Joseph. *Die Tagebücher von Joseph Goebbels.* Ed. Elke Fröhlich. Munich: K. G. Saur, 1987ff.

Gross, Walter. "Vorwort." In Adolf Hitler, *Volk und Rasse.* 3rd ed. Munich: Zentralverlag der NSDAP, Franz Eher Nachf., n.d.

Gruber, Max von. *Ursachen und Bekämpfung des Geburtenrückgangs im Deutschen Reich.* 3rd ed. Munich: J. F. Lehmann, 1914.

Günther, Hans F. K. *Mein Eindruck von Adolf Hitler.* Pähl, F. von Bebenburg, 1969.

————. *Rassenkunde des deutschen Volkes.* 3rd ed. Munich: J. F. Lehmanns Verlag, 1923.

Gütt, Arthur, Ernst Rüdin, and Falk Ruttke, *Gesetz zur Verhütung erbkranken Nachwuchses vom 14. Juli 1933.* Munich: J. F. Lehmanns Verlag, 1934.

Haeckel, Ernst. *Freie Wissenschaft und freie Lehre.* Stuttgart: E. Schweizerbart'sche Verlagshandlung, 1878.

————. *Die Lebenswunder: Gemeinverständliche Studien über Biologische Philosophie.* Stuttgart: Alfred Kröner, 1904.

————. *Natürliche Schöpfungsgeschichte.* Berlin: Georg Reimer, 1868.

————. *Natürliche Schöpfungsgeschichte,* 2nd ed. Berlin: Georg Reimer, 1870.

————. *Über Arbeitstheilung in Natur- und Menschenleben.* Berlin: C. G. Lüderitz'sche Verlagsbuchhandlung, 1869.

————. "Die Wissenschaft und der Umsturz," *Die Zukunft* 10 (1895): 197–206.

Halder, Franz. *Kriegstagebuch.* Ed. Hans-Adolf Jacobsen. Stuttgart: W. Kohlhammer Verlag, 1963.

Handwörterbuch der Naturwissenschaften. Jena: Gustav Fischer, 1912–1913.

Hellwald, Friedrich. *Culturgeschichte in ihrer natürlichen Entwicklung bis zur Gegenwart.* Augsburg: Lampart, 1875.

Hentschel, Willibald. *Varuna: Das Gesetz des aufsteigenden und sinkenden Lebens in der Geschichte.* Leipzig: Theodor Fritsch, 1907.

Hitler, Adolf. *Adolf Hitler spricht: Ein Lexikon des Nationalsozialismus.* Leipzig: R. Kittler Verlag, 1934.

————. *"Es spricht der Führer": 7 exemplarische Hitler-Reden.* Ed. Hildegard von Kotze and Helmut Krausnick. Gütersloh: Sigbert Mohn Verlag, 1966.

————. *Hitler. Reden, Schriften, Anordnungen: Februar 1925 bis Januar 1933.* Ed. Insitut für Zeitgeschichte. 5 vols. in 12 parts. 1992–1998.

————. *Hitler. Sämtliche Aufzeichnungen, 1905–1924.* Ed. Eberhard Jäckel. Stuttgart: Deutsche Verlags-Anstalt, 1980.

————. *Hitler's Second Book: The Unpublished Sequel to* Mein Kampf. Ed. Gerhard Weinberg. New York: Enigma Books, 2003.

————. *Hitlers Zweites Buch: Ein Dokument aus dem Jahr 1928.* Ed. Gerhard L. Weinberg. Stuttgart: Deutsche Verlags-Anstalt, 1961.

————. *Mein Kampf.* Trans. Ralph Manheim. Boston: Houghton Mifflin, 1943.

————. *Mein Kampf.* 2 vols. in 1. Munich: NSDAP, 1943.

———— *Monologe im Führerhauptquartier 1941–1944. Die Aufzeichnungen Heinrich Heims.* Ed. Werner Jochmann. Hamburg: Albrecht Knaus, 1980.

————. *Reden an die deutsche Frau.* Berlin: Schadenverhütung Verlagsgesellschaft, n.d.

————. *The Speeches of Adolf Hitler, April 1922–August 1939.* Ed. Norman H. Baynes. 2 vols. Oxford: Oxford University Press, 1942.

————. *Vortrag Adolf Hitlers von westdeutscher Wirtschaftlern im Industrie-Klub zu Dusseldorf am 27. Januar 1932.* Munich: Frz. Eher Nachf., n.d.

Hossbach Memorandum, at www.yale.edu/lawweb/avalon/imt/hossbach.htm, accessed October 3, 2001.

Junge, Traudl. *Bis zur letzten Stunde: Hitlers Sekretärin erzählt ihr Leben.* Munich: Claassen Verlag, 2002.

Kaiser, Jochen-Christoph, Kurt Nowak, and Michael Schwartz, eds. *Eugenik Sterilisation "Euthanasie": Politische Biologie in Deutschland 1895–1945: Eine Dokumentation.* Berlin: Buchverlag Union, 1992.

Kersten, Felix. *The Kersten Memoirs, 1940–1945.* Trans. Constantine Fitzgibbon and James Oliver. New York: Macmillan, 1957.

Kubizek, August. *The Young Hitler I Knew.* Trans. E. V. Anderson. Boston: Houghton Mifflin, 1955.

Lenz, Fritz. *Die Rasse als Wertprinzip, Zur Erneuerung der Ethik.* Munich: J. F. Lehmann, 1933.

Linge, Heinz. *Bis zum Untergang: Als Chef des Persönlichen Dienstes bei Hitler.* 2nd ed. Munich: Herbig, 1980.

Maser, Werner. *Hitlers Briefe und Notizen. Sein Weltbild in handschriftlichen Dokumenten.* Düsseldorf: Econ Verlag, 1973.

Moll, Martin, ed. *"Führer-Erlasse" 1939–1945.* Stuttgart: Franz Steiner Verlag, 1997.

Müller, Karl Alexander von. *Im Wandel einer Welt: Erinnerungen 1919–1932.* Munich: Süddeutscher Verlag, 1966.

Müller, Reinhard. "Hitlers Rede vor der Reichswehrführung 1933: Eine neue Moskauer Überlieferung," *Mittelweg* 36, 1 (2001): 73–90.

Nazi Conspiracy and Aggression. United States Office of Chief of Counsel for the Prosecution of Axis Criminality. United States Government Printing Office. Washington, DC. 1946.

Noakes, J., and G. Pridham, *Nazism 1919–1945: A Documentary Reader.* 4 vols. Exeter: University of Exeter Press, 1995–2000.

Opfer der Vergangenheit [Nazi film], 1937.

Phelps, Reginald H. "Hitler als Parteiredner," *Vierteljahrshefte für Zeitgeschichte* 11 (1963): 274–330.

Picker, Henry, ed. *Hitlers Tischgespräche im Führerhauptquartier.* Frankfurt: Ullstein, 1989.

Ploetz, Alfred. "Neo-Malthusianism and Race Hygiene." In *Problems in Eugenics: Report of Proceedings of the First International Eugenics Congress.* Vol. 2. London, 1913. Pp. 183ff.

Preyer, Wilhelm. *Die Concurrenz in der Natur.* Breslau: S. Schottlaender, 1882.

———. *Der Kampf um das Dasein.* Bonn: Weber, 1869.

Rassenpolitik. Berlin: Der Reichsführer SS, SS-Hauptamt, n.d.

Ratzel, Friedrich. *Erdenmacht und Völkerschicksal. Eine Auswahl aus seinen Werken.* Ed. Karl Haushofer. Stuttgart: Alfred Kröner, 1940.

Ribbentrop, Joachim. *The Ribbentrop Memoirs.* London: Weidenfeld and Nicolson, 1954.

Rosenberg, Alfred. *Letzte Aufzeichnungen: Nürnberg 1945/46.* 2nd ed. Uelzen: Jomsburg-Verlag, 1996.

Schallmayer, Wilhelm. *Beiträge zu einer Nationalbiologie.* Jena: Hermann Costenoble, 1905.

———. "Rassedienst," *Sexual-Probleme* 7 (1911): 433–443, 534–547.

———. *Vererbung und Auslese im Lebenslauf der Völker. Eine Staatswissenschaftliche Studie auf Grund der neueren Biologie.* Jena: Gustav Fischer, 1903.

Schaumburg-Lippe, Friedrich Christian Prinz zu. *Als die golden Abendsonne: Aus meinen Tagebüchern der Jahre 1933–1937.* Wiesbaden: Limes Verlag, 1971.

Schemann, Ludwig. *Lebensfahrten eines Deutschen.* Leipzig: Erich Matthes, 1925.

———. *Die Rasse in den Geisteswissenschaften: Studien zur Geschichte des Rassengedankens.* Vol. 3: *Die Rassenfragen im Schrifttum der Neuzeit.* 2nd ed. Munich: J. F. Lehmann, 1943.

Schirach, Baldur von. *Ich glaubte an Hitler.* Hamburg: Mosaik Verlag, 1967.

Sebottendorff, Rudolf von. *Bevor Hitler kam: Urkundliches aus der Frühzeit der nationalsozialistischen Bewegung.* Munich: Deukula Verlag, 1933.

Stalin, Joseph V. *Works.* Moscow: Foreign Languages Publishing House, 1955.

Stuckart, Wilhelm, and Hans Globke. *Reichsbürgergesetz vom 15. September 1935, Gesetz zum Schutze des deutschen Blutes und der deutschen Ehre vom 15. September 1935, Gesetz zum Schutze der Erbgesundheit des deutschen Volkes (Ehegesundheitsgesetz) vom 18. Oktober 1935 nebst allen Ausführungsvorschriften und den einschlägigen Gesetzen und Verordnungen.* Munich: C. H. Beck'sche Verlagsbuchhandlung, 1936.

Usadel, Georg. *Zucht und Ordnung: Grundlagen einer nationalsozialistischen Ethik.* 3rd ed. Hamburg: Hanseatische Verlagsanstalt, 1935.

Vogt, Carl. *Vorlesungen über den Menschen, seine Stellung in der Schöpfung und in der Geschichte der Erde.* 2 vols. Giessen, 1863.

Wagener, Otto. *Hitler—Memoirs of a Confidant.* Ed. Henry Ashby Turner. Trans. Ruth Hein. New Haven: Yale University Press, 1985.

Weismann, August. *Aufsätze über Vererbung und Verwandte Biologische Fragen.* Jena: Gustav Fischer, 1892.

Wilhelm, Hans-Heinrich. "Hitlers Ansprache vor Generalen und Offizieren am 26. Mai 1944," *Militärgeschichtliche Mitteilungen* 2 (1976): 123–170.

Wofür kämpfen wir? Berlin: Heerespersonalamt, 1944.

Woltmann, Ludwig. *Die Darwinsche Theorie und der Sozialismus.* Düsseldorf: Hermann Michels Verlag, 1899.

———. *Die Germanen und die Renaissance in Italien.* Leipzig: Thüringische Verlagsanstalt, 1905.

———. *Politische Anthropologie: Eine Untersuchung über den Einfluss der Deszendenztheorie auf die Lehre von der politischen Entwicklung der Völker.* Jena: Eugen Diederichs, 1903.

———. *Werke.* Vol 1: *Politische Anthropologie.* Ed. Otto Reche. Leipzig: Justus Dörner, 1936.

Ziegler, Heinrich Ernst. "Einleitung zu dem Sammelwerke Natur und Staat." In *Natur und Staat.* Vol. 1 (bound with Heinrich Matzat, *Philosophie der Anpassung*). Jena: Gustav Fischer, 1903.

Secondary Sources

Aly, Götz, and Karl Heinz Roth. *The Nazi Census: Identification and Control in the Third Reich* Trans. Edwin Black and Assenka Oksiloff. Philadephia: Temple University Press, 2004.

Auerbach, Hellmuth. "Hitlers politische Lehrjahre und die Münchener Gesellschaft 1919–1923," *Vierteljahrshefte für Zeitgeschichte* 25 (1977): 1–45.

Ayass, Wolfgang. *"Asoziale" im Nationalsozialismus.* Stuttgart: Klett-Cotta, 1995.

Barta, Tony. "Mr. Darwin's Shooters: On Natural Selection and the Naturalizing of Genocide," *Patterns of Prejudice* 39 (2005): 116–137.

Bartov, Omer. *Hitler's Army: Soldiers, Nazis, and War in the Third Reich.* Oxford: Oxford University Press, 1991.

Bauman, Zygmunt. *Modernity and the Holocaust.* Ithaca: Cornell University Press, 1989.

Berenbaum, Michael, ed. *A Mosaic of Victims.* New York: New York University Press, 1990.

Berkhoff, Karel C. *Harvest of Despair: Life and Death in Ukraine under Nazi Rule.* Cambridge, MA: Belknap Press of Harvard University Press, 2004.

Berkowitz, Michael. *The Crime of My Very Existence: Nazism and the Myth of Jewish Criminality.* Berkeley: University of California Press, 2007.

Bessel, Richard. *Nazism and War.* New York: Modern Library, 2004.

Biesold, Horst. *Crying Hands: Eugenics and Deaf People in Nazi Germany.* Trans. William Sayers. Washington, DC: Gallaudet University Press, 1999.

Blackburn, Gilmer. *Education in the Third Reich: Race and History in Nazi Textbooks.* Albany: State University of New York Press, 1985.

Bleker, Johanna, and Norbert Jachertz, eds. *Medizin im "Dritten Reich."* 2nd ed. Cologne: Deutscher Ärzte-Verlag, 1993.

Bock, Gisela. *Zwangssterilisation im Nationalsozialismus: Studien zur Rassenpolitik und Frauenpolitik.* Opladen: Westdeutscher Verlag, 1986.

Bodo, Bela. "The Medical Examination and Biological Selection of University Students in Nazi Germany," *Bulletin of the History of Medicine* 76 (2002): 719–748.

Bracher, Karl Dietrich. *Die Deutsche Diktatur. Entstehung, Struktur, Folgen des Nationalsozialismus.* 7th ed. Cologne: Kiepenheuer and Witsch, 1993.

Breitman, Richard. *The Architect of Genocide: Himmler and the Final Solution.* New York: Alfred A. Knopf, 1991.

Browne, Janet. *Charles Darwin.* Vol. 1: *Voyaging.* New York: Alfred Knopf, 1995.

Browning, Christopher, with contributions by Jürgen Matthäus. *The Origins of the Final Solution: The Evolution of Nazi Jewish Policy, September 1939–March 1942.* Lincoln: University of Nebraska Press, 2004.

Bryant, Chad. *Prague in Black: Nazi Rule and Czech Nationalism.* Cambridge, MA: Harvard University Press, 2007.

Bullock, Alan. *Hitler and Stalin: Parallel Lives.* New York: Alfred Knopf, 1992.

Burleigh, Michael. *Death and Deliverance: Euthanasia in Germany, 1900–1945.* Cambridge: Cambridge University Press, 1994.

———. *The Third Reich: A New History.* New York: Hill and Wang, 2000.

Burleigh, Michael, and Wolfgang Wippermann. *The Racial State: Germany, 1933–1945.* Cambridge: Cambridge University Press, 1991.

Butcher, Barry W. "Darwinism, Social Darwinism and the Australian Aborigines: A Reevaluation." In *Darwin's Laboratory: Evolutionary Theory and Natural History in the Pacific.* Ed. Roy MacLeod and Philip F. Rehbock. Honolulu: University of Hawaii Press, 1994.

Bytwerk, Randall. *Bending Spines: The Propaganda of Nazi Germany and the German Democratic Republic.* East Lansing: Michigan State University Press, 2004.

Chickering, Roger, Stig Förster, and Bernd Greiner, eds. *A World at Total War: Global Conflict and the Politics of Destruction, 1937–1945.* Cambridge: Cambridge University Press, 2005.

Cocks, Geoffrey. "Sick Heil: Self and Illness in Nazi Germany," *Osiris* 22 (2007): 93–115.

Connelly, John. "Nazis and Slavs: From Racial Theory to Racist Practice," *Central European History* 32 (1999): 1–33.

Czarnowski, Gabriele. *Das kontrollierte Paar: Ehe- und Sexualpolitik im Nationalsozialismsus.* Weinheim: Deutscher Studien Verlag, 1991.

Daim, Wilfried. *Der Mann, der Hitler die Ideen Gab: Jörg Lanz von Liebenfels.* 3rd ed. Vienna: Ueberreuter, 1994.

Dawidowicz, Lucy S. *The War against the Jews.* New York: Bantam Books, 1975.

Deadly Medicine: Creating the Master Race. Washington, DC: United States Holocaust Memorial Museum, 2004.

Dickinson, Edward Ross. *The Politics of German Child Welfare from the Empire to the Federal Republic.* Cambridge, MA: Harvard University Press, 1996.

Doeleke, W. "Alfred Ploetz (1860–1940): Sozialdarwinist und Gesellschaftsbiologe," dissertation, University of Frankfurt, 1975.

Dowbiggin, Ian. *A Merciful End: The Euthanasia Movement in Modern America.* Oxford: Oxford University Press, 2003.

Eckart, Wolfgang U., ed. *Man, Medicine, and the State: The Human Body as an Object of Government Sponsored Medical Research in the 20th Century.* Stuttgart: Franz Steiner Verlag, 2006.

Ehrenreich, Eric. *The Nazi Ancestral Proof: Genealogy, Racial Science, and the Final Solution.* Bloomington: Indiana University Press, 2007.

———. "Otmar von Verschuer and the 'Scientific' Legitimization of Nazi Anti-Jewish Policy," *Holocaust and Genocide Studies* 21, 1 (2007): 55–72.

Essner, Cornelia. *Die "Nürnberger Gesetze" oder Die Verwaltung des Rassenwahns 1933–1945.* Paderborn: Ferdinand Schöningh, 2002.

Evans, Richard. *The Coming of the Third Reich.* New York: Penguin, 2004.

————. "In Search of German Social Darwinism: The History and Historiography of a Concept." In *Medicine and Modernity: Public Health and Medical Care in Nineteenth- and Twentieth-Century Germany*. Ed. Manfred Berg and Geoffrey Cocks. Washington, DC: Cambridge University Press for the German Historical Institute, 1997. 55–79.

————. *The Third Reich in Power*. New York: Penguin, 2005.

Fangerau, Heiner. *Etablierung eines rassenhygienischen Standardwerkes 1921–1941: Der Baur-Fischer-Lenz im Spiegel der zeitgenössischen Rezensionsliteratur*. Frankfurt: Peter Lang, 2001.

Fest, Joachim C. *The Face of the Third Reich: Portraits of the Nazi Leadership*. Trans. Michael Bullock. New York: Pantheon, 1970.

————. *Hitler*. Trans. Richard and Clara Winston. New York: Helen and Kurf Wolff, 1974.

Frei, Norbert. "Wie modern war der Nationalsozialismus," *Geschichte und Gesellschaft* 19 (1993): 367–387.

Frewer, Andreas. *Medizin und Moral in Weimarer Republik und Nationalsozialismus: die Zeitschrift "Ethik" unter Emil Abderhalden*. Frankfurt: Campus Verlag, 2000.

Friedlander, Henry. *The Origins of Nazi Genocide: From Euthanasia to the Final Solution*. Chapel Hill: University of North Carolina Press, 1995.

Friedländer, Saul. *Nazi Germany and the Jews*. Vol. 1: *The Years of Persecution, 1933–1939*. New York: Harper Collins, 1997.

————. *The Years of Extermination: Nazi Germany and the Jews, 1939–1945*. New York: Harper Perennial, 2007.

Fritzsche, Peter. *Germans into Nazis*. Cambridge, MA: Harvard University Press, 1998.

Gasman, Daniel. *The Scientific Origins of National Socialism: Social Darwinism in Ernst Haeckel and the German Monist League*. London: MacDonald, 1971.

Gassert, Phillip, and Daniel S. Mattern, eds. *The Hitler Library: A Bibliography*. Westport, Conn.: Greenwood Press, 2001.

Gellately, Robert. *Backing Hitler: Consent and Coercion in Nazi Germany*. Oxford: Oxford University Press, 2001.

Gellately, Robert, and Nathan Stolzfus, eds. *Social Outsiders in Nazi Germany*. Princeton: Princeton University Press, 2001.

Gerlach, Christian. "The Wannsee Conference, the Fate of the German Jews, and Hitler's Decision in Principle to Exterminate All European Jews," *Journal of Modern History* 70 (1998): 759–812.

Germany and the Second World War. Ed. Militärgeschichtliches Forschungsamt, Freiburg. Vol. 1: *The Build-up of German Aggression*. Oxford: Clarendon Press, 1990.

Glover, Jonathan. *Humanity: A Moral History of the Twentieth Century*. New Haven: Yale Nota Bene, 2001.

Goldhagen, Daniel. *Hitler's Willing Executioners: Ordinary Germans and the Holocaust*. New York: Knopf, 1996.

Gregor, Neil, *How to Read Hitler*. New York: W.W. Norton, 2005.

Griffin, Roger. *Modernism and Fascism: The Sense of a Beginning under Mussolini and Hitler*. New York: Palgrave Macmillan, 2007.

Gross, Jan Tomasz. *Polish Society under German Occupation: The Generalgouvernement, 1939–1944*. Princeton: Princeton University Press, 1979.

Grossman, Atina. *Reforming Sex: The German Movement for Birth Control and Abortion Reform, 1920–1950*. Oxford: Oxford University Press, 1995.

Haar, Ingo, and Michael Fahlbusch, eds. *German Scholars and Ethnic Cleansing, 1920–1945*. New York: Berghahn Books, 2005.

Haas, Peter J. *Morality after Auschwitz: The Radical Challenge of the Nazi Ethic*. Philadelphia: Fortress Press, 1988.

Haas, Peter J. "Science and the Determination of the Good," in *Ethics after the Holocaust: Perspectives, Critiques, and Responses*. Ed. John Roth. St. Paul: Paragon House, 1999. 49–89.

Haffner, Sebastian. *The Meaning of Hitler*. Trans. Ewald Osers. Cambridge, MA: Harvard University Press, 1983.

Hamann, Brigitte. *Hitler's Vienna: A Dictator's Apprenticeship*. Trans. Thomas Thornton. New York: Oxford University Press, 1999.

Harten, Hans-Christian. *De-Kulturation und Germanisierung: Die nationalsozialistische Rassen- und Erziehungspolitik in Polen 1939–1945*. Frankfurt: Campus Verlag, 1996.

Hawkins, Mike. *Social Darwinism in European and American Thought, 1860–1945: Nature as Model and Nature as Threat*. Cambridge: Cambridge University Press, 1997.

Herbert, Ulrich. *Fremdarbeiter: Politik und Praxis des "Ausländer-Einsatzes" in der Kriegswirtschaft des Dritten Reiches*. Berlin: J. H. W. Dietz Nachf., 1985.

Herf, Jeffrey. *The Jewish Enemy: Nazi Propaganda during World War II and the Holocaust*. Cambridge, MA: Belknap Press of Harvard University Press, 2006.

———. *Reactionary Modernism: Technology, Culture, and Politics in Weimar and the Third Reich*. Cambridge: Cambridge University Press, 1984.

Hering, Rainer. *Konstruierte Nation: Der Alldeutsche Verband 1890 bis 1939*. Hamburg: Christians, 2003.

Hermand, Jost. *Old Dreams of a New Reich: Volkish Utopias and National Socialism*. Trans. Paul Levesque. Bloomington: Indiana University Press, 1992.

Herzog, Dagmar. *Sex after Fascism: Memory and Morality in Twentieth-Century Germany*. Princeton: Princeton University Press, 2005.

Herzstein, Robert Edwin. *The War that Hitler Won: The Most Infamous Propaganda Campaign in History*. New York: G. P. Putnam's Sons, 1978.

Hipler, Bruno. *Hitlers Lehrmeister: Karl Haushofer als Vater der NS-Ideologie*. St. Ottilien: EOS-Verlag, 1996.

Housden, Martyn. *Hans Frank: Lebensraum and the Holocaust*. Houndmills: Palgrave Macmillan, 2003.

Hutton, Christopher. *Race and the Third Reich: Linguistics, Racial Anthropology and Genetics in the Dialectic of Volk*. Cambridge, UK: Polity, 2005.

Jäckel, Eberhard. *Hitler's Weltanschauung: A Blueprint for Power*. Trans. Herbert Arnold Middleton, CT: Wesleyan University Press, 1972.

Jacobsen, Hans-Adolf. *Karl Haushofer: Leben und Werk*. Vol. 1: *Lebensweg 1869–1946 und ausgewählte Texte zur Geopolitik*. Boppard am Rhein: Harald Boldt Verlag, 1979.

Kay, Alex J. *Exploitation, Resettlement, Mass Murder: Political and Economic Planning for German Occupation Policy in the Soviet Union, 1940–1941*. New York: Berghahn Books, 2006.

Kellogg, Michael. *The Russian Roots of Nazism: White Émigrés and the Making of National Socialism, 1917–1945*. Cambridge: Cambridge University Press, 2005.

Kemp, N. D. A. *'Merciful Release': The History of the British Euthanasia Movement*. Manchester: Manchester University Press, 2002.

Kershaw, Ian. *Hitler*. 2 vols. New York: Norton, 1998–2000.

———. *The Hitler Myth*. New York: Clarendon Press of Oxford University Press, 1987.

King, Richard H. and Dan Stone. *Hannah Arendt and the Uses of History: Imperialism, Nation, Race, and Genocide*. New York: Berghahn Books, 2007.

Klee, Ernst. *Deutsche Medizin im Dritten Reich: Karrieren vor und nach 1945*. Frankfurt: S. Fischer, 2001.

———. *"Euthanasie" im NS-Staat: Die "Vernichtung lebensunwerten Lebens."* Frankfurt: Fischer Taschenbuch, 1985.

Koonz, Claudia. *Mothers in the Fatherland*. New York: St. Martin's Press, 1987.

———. *The Nazi Conscience*. Cambridge, MA: Harvard University Press, 2003.

Kühl, Stefan. *The Nazi Connection: Eugenics, American Racism, and German National Socialism.* Oxford: Oxford University Press, 1994.

———. "The Relationship between Eugenics and the So-Called 'Euthanasia Action' in Nazi Germany: A Eugenically Motivated Peace Policy and the Killing of the Mentally Handicapped during the Second World War." In *Science and the Third Reich*. Ed. Margit Szöllösi-Janze. Oxford: Berg, 2001. 185–210.

Lange, Karl. "Der Terminus 'Lebensraum' in Hitlers *Mein Kampf*," *Vierteljahrshefte für Zeitgeschichte* 13 (1965): 426–437.

Large, David Clay. "'Darktown Parade': African Americans in the Berlin Olympics of 1936," *Historically Speaking* 9, 2 (November/December 2007).

Lilienthal, Georg. "Die jüdischen 'Rassenmerkmale': Zur Geschichte der Anthropologie der Juden," *Medizinhistorisches Journal* 28, 2/3 (1993): 173–198.

———. *Der Lebensborn e. V.: Ein Instrument nationalsozialistischer Rassenpolitik.* Frankfurt a.M.: Fischer Taschenbuch Verlag, 1993.

Longerich, Peter. *The Unwritten Order: Hitler's Role in the Final Solution.* Tempus, 2003.

Lösch, Niels. *Rasse als Konstrukt: Leben und Werk Eugen Fischers.* Frankfurt a.M.: Lang, 1997.

Löwenberg, Dieter. *Willibald Hentschel (1858–1947): Seine Pläne zur Menschenzüchtung, sein Biologismus und Antisemitismus.* Dissertation, University of Mainz, 1978.

Lower, Wendy. *Nazi Empire-Building and the Holocaust in Ukraine.* Chapel Hill: University of North Carolina Press, 2005.

———. "A New Ordering of Space and Race: Nazi Colonial Dreams in Zhytomyr, Ukraine, 1941–1944," *German Studies Review* 25 (2002): 227–254.

Lukacs, John. *The Hitler of History.* New York: Vintage, 1997.

Mai, Uwe. *"Rasse und Raum": Agrarpolitik, Sozial-und Raumplanung im NS-Staat.* Paderborn: Ferdinand Schöningh, 2002.

Majer, Diemut. *"Non-Germans" under the Third Reich: The Nazi Judicial and Administrative System in Germany and Occupied Eastern Europe, with Special Regard to Occupied Poland, 1939–1945.* Trans. Peter Thomas Hill, Edward Vance Humphrey, and Brian Levin. Baltimore: Johns Hopkins University Press, 2003.

Maser, Werner. *Adolf Hitler: Legende, Mythos, Wirklichkeit.* Munich: Bechtle, 1971.

———. *Adolf Hitler, Mein Kampf: Geschichte, Auszüge, Kommentare.* 6th ed. Esslingen: Bechtle, 1981.

Massin, Benoit. "From Virchow to Fischer: Physical Anthropology and 'Modern Race Theories' in Wilhelmine Germany." In *Volksgeist as Method and Ethic.* Ed. George W. Stocking. Madison: University of Wisconsin Press, 1996. 79–154.

Mastny, Vojtech. *The Czechs under Nazi Rule: The Failure of National Resistance, 1939–1942.* New York: Columbia University Press, 1971.

Megargee, Geoffrey. *War of Annihilation: Combat and Genocide on the Eastern Front, 1941.* Lanham, MD: Rowman and Littlefield, 2006.

Meinel, Christoph and Peter Voswinckel, eds. *Medizin, Naturwissenschaft, Technik und Nationalsozialismus: Kontinuitäten und Diskontinuitäten.* Stuttgart: Verlag für Geschichte der Naturwissenschaft und der Technik, 1994.

Mosse, George L. *Nationalism and Sexuality: Respectability and Abnormal Sexuality in Modern Europe.* New York: Howard Fertig, 1985.

———. *Nazi Culture: Intellectual, Cultural, and Social Life in the Third Reich.* New York: Grosset and Dunlap, 1966.

Mouton, Michelle. *From Nurturing the Nation to Purifying the Volk: Weimar and Nazi Family Policy, 1918–1945.* Cambridge: Cambridge University Press, 2007.

Müller-Hill, Benno. *Murderous Science: Elimination by Scientific Selection of Jews, Gypsies, and Others, Germany, 1933–1945.* Trans. George R. Fraser. Oxford: Oxford University Press, 1988.

Noakes, Jeremy. "The Development of Nazi Policy towards the German-Jewish 'Mischlinge' 1933–1945," *Leo Baeck Institute Yearbook* 34 (1989): 291–354.

Overy, Richard. *The Dictators: Hitler's Germany and Stalin's Russia.* New York: W. W. Norton, 2004.

———. *Interrogations: The Nazi Elite in Allied Hands, 1945.* New York: Viking, 2001.

———. *War and Economy in the Third Reich.* Oxford: Clarendon Press, 1994.

Pine, Lisa. *Nazi Family Policy, 1933–1945.* Oxford: Berg, 1997.

Plöckinger, Othmar. *Geschichte eines Buches: Adolf Hitlers "Mein Kampf" 1922–1945.* Munich: R. Oldernbourg, 2006.

Pommerin, Rainer. *Sterilisierung der Rheinlandbastarde. Das Schicksal einer farbigen deutschen Minderheit 1918–1937.* Düsseldorf: Droste Verlag, 1979.

Prinz, Michael and Rainer Zitelmann, eds. *Nationalsozialismus und Modernisierung.* Darmstadt: Wissenschaftliche Buchgesellschaft, 1991.

Proctor, Robert. *Racial Hygiene: Medicine under the Nazis.* Cambridge, MA: Harvard University Press, 1988. 414.

Przyrembel, Alexandra. *'Rassenschande': Reinheitsmythos und Vernichtungslegitimation im Nationalsozialismus.* Göttingen: Vandenhoek und Ruprecht, 2003.

Rich, Norman. *Hitler's War Aims: Ideology, the Nazi State, and the Course of Expansionism.* New York: Norton, 1973.

Rosenbaum, Ron. *Explaining Hitler.* New York: Random House, 1998.

Rossino, Alexander B. *Hitler Strikes Poland: Blitzkrieg, Ideology, and Atrocity.* Lawrence: University Press of Kansas, 2003.

Rubenstein, Richard. "Modernization and the Politics of Extermination." In *A Mosaic of Victims.* Ed. Michael Berenbaum. New York: New York University Press, 1990.

Rutherford, Phillip. *Prelude to the Final Solution: The Nazi Program for Deporting Ethnic Poles, 1939–1941.* Lawrence: University Press of Kansas, 2007.

Sandmann, Jürgen. *Der Bruch mit der humanitären Tradition. Die Biologisierung der Ethik bei Ernst Haeckel und anderen Darwinisten seiner Zeit.* Stuttgart: Gustav Fischer Verlag, 1990.

Schafft, Gretchen E. *From Racism to Genocide: Anthropology in the Third Reich.* Urbana: University of Illinois Press, 2004.

Schaller, Helmut. *Der Nationalsozialismus und die slawische Welt.* Regensburg: Verlag Friedrich Pustet, 2002.

Scheck, Raffael. *Hitler's African Victims: The German Army Massacres of Black French Soldiers in 1940.* Cambridge: Cambridge University Press, 2006.

Schmidt, Ulf. *Medical Films, Ethics, and Euthanasia in Nazi Germany: The History of Medical Research and Teaching Films of the Reich Office for Educational Films/Reich Institute for Films in Science and Education, 1933–1945.* Husum: Matthiesen Verlag, 2002.

Schmuhl, Hans-Walter. *Grenzüberschreitungen: Das Kaiser-Wilhelm-Institut für Anthropologie, menschliche Erblehre und Eugenik 1927–1945.* Göttingen: Wallstein Verlag, 2005.

———. *Rassenhygiene, Nationalsozialismus, Euthanasie. Von der Verhütung zur Vernichtung 'lebensunwerten Lebens' 1890–1945.* Göttingen: Vandenhoek und Ruprecht, 1987.

———, ed. *Rassenforschung an Kaiser-Wilhelm-Instituten vor und nach 1933.* Göttingen: Wallstein Verlag, 2003.

Smith, Woodruff. *Origins of Nazi Imperialism.* New York: Oxford University Press, 1986.

Spotts, Frederic. *Hitler and the Power of Aesthetics.* Woodstock: Overlook Press, 2003.

Staudinger, Hans. *The Inner Nazi: A Critical Analysis of Mein Kampf.* Baton Rouge: Louisiana State University Press, 1981.

Steigmann-Gall, Richard. *The Holy Reich: Nazi Conceptions of Christianity, 1919–1945* Cambridge: Cambridge University Press, 2003.

Steinweis, Alan. *Studying the Jew: Scholarly Antisemitism in Nazi Germany.* Cambridge, MA: Harvard University Press, 2006.

Stephenson, Jill. *Women in Nazi Society.* New York: Barnes and Noble, 1975.

Stern, Fritz. *Politics of Cultural Despair.* Garden City, NY: Anchor Books, 1965.

Stoakes, Geoffrey. *Hitler and the Quest for World Dominion.* Leamington Spa, UK: Berg, 1986.

Stöckel, Sigrid, ed. *Die "rechte Nation" und Ihr Verleger: Politik und Popularisierung im J. F. Lehmanns Verlag 1890–1979.* Berlin: Lehmanns, 2002.

Tent, James F. *In the Shadow of the Holocaust: Nazi Persecution of Jewish-Christian Germans.* Lawrence: University of Kansas Press, 2003.

Thomann, Klaus-Dieter and Werner Friedrich Kümmel. "Naturwissenschaft, Kapital und Weltanschauung: Das Kruppsche Preisausschreiben und der Sozialdarwinismus," *Medizinhistorisches Journal* 30 (1995): 99–143, 205–243.

Turda, Marius and Paul Weindling. *Blood and Homeland: Eugenics and Racial Nationalism in Central and Southeast Europe, 1900–1940.* Budapest: Central European University Press, 2006.

Walkenhorst, Peter. *Nation—Volk—Rasse: Radikaler Nationalismus im Deutschen Kaiserreich 1890–1914.* Göttingen: Vandenhoeck und Ruprecht, 2007.

Weber, Matthias. *Ernst Rüdin: Eine kritische Biographie.* Berlin: Springer, 1993.

Weikart, Richard. "Darwinism and Death: Devaluing Human Life in Germany, 1860–1920," *Journal of the History of Ideas* 63 (2002): 323–344.

———. *From Darwin to Hitler: Evolutionary Ethics, Eugenics, and Racism in Germany.* New York: Palgrave Macmillan, 2004.

———. "The Impact of Social Darwinism on Anti-Semitic Ideology in Germany and Austria, 1860–1945." In *Jewish Tradition and the Challenge of Evolution.* Ed. Geoffrey Cantor and Mark Swetlitz. Chicago: University of Chicago Press, 2006. 93–115.

———. "Laissez-Faire Social Darwinism and Individualist Competition in Darwin and Huxley," *The European Legacy* 3 (1998): 17–30.

———. "The Origins of Social Darwinism in Germany, 1859–1895," *Journal of the History of Ideas* 54 (1993): 469–488.

———. "Progress through Racial Extermination: Social Darwinism, Eugenics, and Pacifism in Germany, 1860–1918," *German Studies Review* 26 (2003): 273–294.

———. "A Recently Discovered Darwin Letter on Social Darwinism," *Isis* 86 (1995): 609–611.

———. *Socialist Darwinism: Evolution in German Socialist Thought from Marx to Bernstein.* San Francisco: International Scholars Publications, 1999.

———. "Was Darwin or Spencer the Father of Laissez-Faire Social Darwinism?" *Journal of Economic Behavior and Organization* (special issue on social Darwinism) (forthcoming, 2008).

Weinberg, Gerhard. *The Foreign Policy of Hitler's Germany.* Vol. 1: *Diplomatic Revolution in Europe, 1933–36.* Chicago: University of Chicago Press, 1970.

Weindling, Paul. *Health, Race and German Politics between National Unification and Nazism, 1870–1945.* Cambridge: Cambridge University Press, 1989.

Weingart, Peter, Jürgen Kroll, and Kurt Bayertz. *Rasse, Blut, und Gene. Geschichte der Eugenik und Rassenhygiene in Deutschland.* Frankfurt: Suhrkamp, 1988.

Weiss, Sheila Faith. *Race Hygiene and National Efficiency: The Eugenics of Wilhelm Schallmayer.* Berkeley: University of California Press, 1987.

Welch, David. *The Third Reich: Politics and Propaganda.* 2nd ed. London: Routledge, 2002.

Westermann, Edward. *Hitler's Police Battalions: Enforcing Racial War in the East.* Lawrence: University Press of Kansas, 2005.

Wetzell, Richard F. *Inventing the Criminal: A History of German Criminology, 1880–1945*. Chapel Hill: University of North Carolina Press, 2000.

Winter, Bettina, ed. *"Verlegt nach Hadamar": Die Geschichte einer NS-"Euthanasie"-Anstalt*. Kassel: Landeswolfahrtsverband Hessen, 1991.

Wippermann, Wolfgang. *Der consequente Wahn. Ideologie und Politik Adolf Hitlers*. Gütersloh: Bertelsmann, 1989.

Wistrich, Robert S. *Hitler and the Holocaust*. New York: Modern Library, 2003.

Zimmerman, Andrew. *Anthropology and Antihumanism in Imperial Germany*. Chicago: University of Chicago Press, 2001.

Zitelmann, Rainer. *Hitler: Selbstverständnis eines Revolutionärs*. Hamburg: Berg, 1987.

Zmarzlik, Hans-Günter. "Der Sozialdarwinismus in Deutschland als geschichtliches Problem," *Vierteljahrshefte für Zeitgeschichte* 11 (1963): 246–273.

INDEX

Note: The following terms are not indexed, since they occur too often throughout the book: Hitler, Nazism, evolution, ethics, morality, and progress.